# The Easiest Flowers to Grow

Created and designed by
the editorial staff of
ORTHO BOOKS

Project Editor
**Janet Goldenberg**

Writer and Photographer
**Derek Fell**

Illustrators
**Cyndie Clark-Huegel**
**Ron Hildebrand**

Designer
**Gary Hespenheide**

# Ortho Books

**Publisher**
Edward A. Evans

**Editorial Director**
Christine Jordan

**Production Director**
Ernie S. Tasaki

**Managing Editors**
Michael D. Smith
Sally W. Smith

**System Manager**
Linda M. Bouchard

**National Sales Manager**
J. D. Gillis

**National Accounts Manager—
Book Trade**
Paul D. Wiedemann

**Marketing Specialist**
Dennis M. Castle

**Distribution Specialist**
Barbara F. Steadham

**Operations Assistant**
Georgiann Wright

**Administrative Assistant**
Francine Lorentz-Olson

**Technical Consultant**
J. A. Crozier, Jr., Ph.D.

Address all inquiries to:
Ortho Books
Box 5006
San Ramon, CA 94583-0906

10  11  12  13  14  15
94  95  96  97  98  99

ISBN 0-89721-220-7
Library of Congress Catalog Card
Number 90-80075

THE SOLARIS GROUP
2527 Camino Ramon
San Ramon, CA 94583-0906

# Acknowledgments

**Manuscript Consultant**
Steven Still, Ph.D.

**Photo Editor**
Sarah Bendersky

**Copy Chief**
Melinda E. Levine

**Editorial Coordinator**
Cass Dempsey

**Copyeditor**
Hazel White

**Proofreader**
Deborah Bruner

**Indexer**
Elinor Lindheimer

**Editorial Assistants**
Nancy McCune
John Parr

**Composition by**
Laurie A. Steele

**Layout and Production by**
Studio 165

**Separations by**
Color Tech Corporation

**Lithographed in the USA by**
Webcrafters, Inc.

**Additional Photographers**
Names of photographers are followed by the page
numbers on which their work appears. R=right,
C=center, L=left, T=top, B=bottom.

William C. Aplin: 104L

John Blaustein: 23

Josephine Coatsworth: 38, 43T, 45B, 48T, 68T, 82L,
    107L

Priscilla Connell, Photo/NATS: 64

Kennon Cooke, Valan: 17

Chris Gotman, Valan: 25

Saxon Holt: 27T, 48C

Michael Landis: 31, 43B, 49L

Michael McKinley: 18, 45T, 46

James D. McNair: 62

A. Millard: 29

Julie O'Neil, Photo/NATS: 60

Ortho Information Services: 34T, 40, 49R, 84C, 89R,
    90C

Pam Peirce: 59

Ann Reilly, Photo/NATS: 85L

David M. Stone, Photo/NATS: 68T

Judith Tankard, Photo/NATS: 39

**Front Cover**
Red zinnias and white Madagascar periwinkles

**Back Cover**
**Top left:** 'Sunshine' gazania

**Top right:** A saucer magnolia blooms over an azalea.

**Bottom left:** Forsythia

**Bottom right:** Impatiens

**Title Page**
Siberian iris

# The Easiest Flowers to Grow

# Easy Does It

*Don't let appearances fool you: Hard work is not necessary for a beautiful flower garden. All you need are the right plants, a few minutes each day to maintain them, and lots of time to enjoy them.*

This is a book about making a flower garden easy in every sense of the word—easy to plant, easy to care for, and easy to appreciate. Easy also means trouble free—resistant to pests, diseases, and environmental stresses such as heat and drought.

There is simply no reason not to have beautiful flowers enhancing your home and its surroundings, no matter how little time or space you have. Once planted, a small garden of less than 500 square feet should require no more than a few minutes of your time each day to produce nonstop color from spring until autumn. If space is even more limited, you'll find that many flowering plants will grow contentedly on a deck or terrace, in window boxes, hanging baskets, and pots; or along a wall or fence as decorative vines. Many of the plants featured here not only look good in the garden but have long, strong stems and wilt-resistant flowers suitable for cutting.

In these pages you will find simple steps to a colorful, flower-filled garden, along with recommendations for the easiest annuals, perennials, shrubs, and small trees for carefree garden color. All of the plants described here are available at garden centers throughout North America or through mail-order sources such as those listed on page 109. As you succeed with familiar varieties, take time to experiment with new ones. You'll find that as your gardening experience grows, so will your confidence. Indeed, you'll wonder why you ever went without a flower garden.

*In this easy-care front yard, wooden landscape ties frame colorful displays of marigolds, zinnias, scarlet sage, and other annuals.*

*Blue and white salvia, red crested cockscomb, and yellow gazania make good companions in this carefree informal garden.*

## YOU CAN SUCCEED

Many aspiring flower gardeners have the best of intentions, but when the time comes to plant they fail to follow through—it rains, or guests appear, or gardening suddenly seems like a chore. Or, having planted, they are daunted by weeds, pests, or drought. As with all worthwhile endeavors, persistence is the key. Follow the recommendations in this book, and your reward will be a garden full of carefree flowers and the satisfaction of developing a "green thumb."

Perhaps the best advice for new gardeners is to start small. A small garden that is easy to maintain will bring far more satisfaction than a large one. It is tempting to order dozens of flowers from a glamorous seed catalog, but if you select just a few easy-to-grow, long-flowering varieties to start with, success will come far more easily. With your newfound confidence, you can work toward a bigger, more colorful garden the following year.

Proper soil preparation is vital. Without well-tended soil, even the best seeds and plants may produce poor results. The first step is to turn over all the soil in the planting area to the depth of a garden spade (about 1 foot), then to rake it carefully to break up soil particles. Add a soil conditioner such as compost or peat moss and, if you know your soil is acid, a neutralizing agent such as garden lime. If you use peat moss or wood products, which lack nitrogen, supplement them with a commercial fertilizer. A formula with a little more nitrogen than phosphorus and potassium, such as 15-10-10 (containing 15 percent nitrogen, 10 percent phosphorus, and 10 percent potassium), works well for most annuals and perennials; formulations with a higher ratio of nitrogen, such as 10-5-5, are best for most shrubs and trees.

Don't plant too early. There's nothing more discouraging than spending a lot of time planting tender flowers such as impatiens, begonias, or petunias, only to have them wiped out by an unexpected late frost. It is actually beneficial to delay planting until well after the last expected frost date. If planted too early, many tender annuals will reach maturity when days start to get hot and humid. This heat stress weakens them and they never recover. Flowers planted later will have the energy to survive heat stress and, as cooler conditions return, will continue blooming to finish the season with a flourish.

Once you have your garden planted, don't neglect it. Water established plantings at least once each week during dry periods, preferably with a watering can or drip feeder that applies water directly to the soil. Pull weeds as soon as they appear. Weeds flourish on neglect, and once they have a stranglehold the job of eradicating them can be difficult—especially when summer days turn hot and you lose the incentive to work outdoors. Removing flowers as soon as they fade will stimulate many plants to bloom longer and more prolifically. If you tidy the garden a bit each day, run a hoe between plants now and then to loosen the soil, and keep an eye open for signs of pest damage, you'll find that your garden almost takes care of itself.

*Flowering azalea and magnolia provide a colorful balance of texture, form, and height.*

## PLANTING FOR SUCCESS

The ideal easy flower garden provides what Connecticut gardener Ruth Levitan calls "a big bang for the buck" (see the description of her intensively planted flower garden on page 10). In other words, it provides bold floral displays that last a long time or come back reliably each year. Such gardens rely on three categories of plants for their bold impact and carefree performance: annuals for rapid, summer-long flowering; perennials for seasonal color and year-to-year continuity; and shrubs and small trees for height and substance augmented by brief bursts of color.

### Abundant Annuals

The quickest way to a carefree flower garden is to plant mostly annuals. Although annuals live only one year, they compensate for their brief existence with rapid and prolific flowering. Many annuals will bloom within six weeks of sowing seeds, compared with a year for most perennials. Some, such as wax begonias, impatiens, and marigolds, will continue blooming for ten weeks or more.

Young seedlings of many annuals are widely available at garden centers, ready to provide instant color if you don't want to fuss with seeds. These ready-to-bloom plants are inexpensive; a flat of 24 can cost less than a single ready-to-bloom perennial in a gallon container. What's more, most annuals need only minimal fertilizing, and the seeds of some (such as calendulas, poppies, and cornflowers) will resow themselves year after year, giving rise to many generations of flowers.

*Planting annuals is the quickest way to a colorful garden. Here, multicolored zinnias combine well with blue salvia and pink and white spider-flowers. The tall pink lythrum is a perennial.*

Annuals are classified as either warm season (tender) or cool season (hardy). Warm-season annuals, such as marigolds, are damaged by frost and flower best during warm, sunny weather. Cool-season annuals, such as snapdragons, usually tolerate mild frosts and bloom best during spring or fall.

The range of flowering annuals available today is so extensive and the cost so economical that you can easily have a beautiful, long-lasting floral display using annuals alone.

### Durable Perennials

Perennials, such as irises and lilies, add sophisticated beauty to a garden. Although most bloom for only a few weeks, they contribute a range of colors, textures, and forms that annuals cannot match. Moreover, their tenacious root systems enable many of them to survive cold winters and live for years.

Most perennials are sold singly in containers and are more expensive than annuals, but their longevity makes them cost efficient. Although many perennials take two years to flower from seed, a year-old container-grown plant will begin to flower immediately. However, use perennials sparingly: Many will grow larger than annuals and can easily overwhelm them in your garden.

Biennials, a category of perennials, live just two years; these are normally planted one summer to flower the next. Most perennials reproduce from seed or by root division. One category of perennials—flowering bulbs—reproduces through a swollen underground food store that splits into miniature replicas of itself to form new plants.

Perennials, like annuals, are classified as hardy or tender. A hardy perennial will survive freezing and a tender one usually will not.

### Sturdy Shrubs and Trees

Thanks to their rigid woody stems, shrubs and small trees carry color above the height of annuals and perennials. The distinction between shrubs and trees is not always sharp, but in general shrubs are short with multiple stems, whereas trees have a single trunk and grow tall—usually 15 feet or higher.

Once they are established, these sturdy plants demand little attention beyond occasional pruning for a pleasing shape, and a yearly application of fertilizer. Small trees give the garden a skyline profile; shrubs fill the gap between flowering trees and lower plantings of annuals and perennials. Although their flowering is usually brief, shrubs and trees form a backdrop against which other plants can shine.

## A NOTE ABOUT NOMENCLATURE

The descriptions in this book's "Plant Selection Guide," and in many seed and plant catalogs, refer to plants by botanical name rather than by common name—for example, *Antirrhinum majus* rather than snapdragon. Although common names are widely used and easy to remember, they sometimes vary from place to place and may even refer to more than one plant—potentially causing confusion. When shopping for plants, it's best to ask for them by botanical name to ensure that you get exactly what you want.

The shortest botanical names consist of two words, such as *Cornus florida*, set in italics. The first word, *Cornus*, identifies the plant's genus (group of related species), in this case dogwood. The second word, *florida*, identifies the species within that genus—here flowering dogwood, a white-flowered native of the eastern seaboard of North America.

Within a species there can be additional variation—occurring in the wild or under cultivation—so that a third word is often added. For example, *Cornus florida* var. *rubra* is a pink-flowering dogwood variety that occurs naturally as a mutation of the white kind.

If a variety has occurred under cultivation, it is called a cultivar, short for "cultivated variety." Cultivar names are usually capitalized in single quotation marks—as in *Cornus florida* 'Fragrant Cloud', an especially fragrant white-flowered cultivar named by the nurseryman who discovered and propagated it. However, the word *cultivar* is used mostly by botanists; in gardening circles the term *variety* is generally used to describe either a variety or a cultivar.

If a plant is the result of hybridizing, its botanical name will usually include the symbol "×," as in *Caryopteris* × *clandonensis*. A hybrid can be the result of a cross between two genera, two species, or two hybrids, or between a species and a hybrid.

*The Levitans'
suburban woodland
features spring-
flowering coralbells,
grapehyacinth, and
daffodils, which bloom
before deciduous trees
shade the garden.
Several trees have
been removed from
the center to admit
more light.*

## RUTH LEVITAN'S FLOWER-FILLED PARADISE

Connecticut gardener Ruth Levitan and her husband, Jim, own what *Architectural Digest* has described as "the most beautiful acre in all America." A woodland garden on a suburban lot, it erupts each spring in an extravaganza of flowering perennials, bulbs, wildflowers, and shrubs. Along meandering grass paths and at the edges of clearings, flowering dogwoods light up the overhead canopy of spring foliage and azaleas glow from the greenery. The woodland floor is a colorful sea of flowering annuals, perennials, and bulbs.

One secret of the garden's success is Mrs. Levitan's ruthless policy of discarding any flower that fails to produce abundant color with little effort and no pampering. "Unlike those gardeners who seek out the challenging and the exotic, I like a big bang for my buck," says Mrs. Levitan. "My ideal plant is covered with large and beautiful flowers during its prolonged blooming season and has attractive foliage the rest of the time. It also has a taste

repugnant to insects, does not get leggy without full sun, or droop from infrequent watering." She praises false-indigo (*Baptisia australis*), black-eyed-susan (*Rudbeckia hirta*), purple coneflower (*Echinacea purpurea*), and hybrid daylily (*Hemerocallis* hybrids). She promptly dispatches malingerers and victims of animal or insect pests to the compost pile, and tries something different. "Shape up or ship out" is her philosophy for a riot of color.

The idea for the garden began some thirty years ago, when the Levitans proudly surveyed their first homesite. The acre of wooded hillside was brilliant with the reds and golds of a Connecticut autumn—but choked with brambles and poison-ivy. The soil, though fertile with leaf mold, contained many boulders and rocks, and the beautiful canopy of deciduous trees obstructed the sunlight. Yet another limitation was the need to rely on precious well water for irrigation. Working within these restrictions the Levitans decided to create a spring garden, where the majority of flowers would bloom before the trees were in full leaf and where flowering would reach its climax before summer drought took its toll.

The rocks became walls for defining boundaries and beds, or foundations for terraces. Several trees were removed from the garden's central area to let in more light for flowering plants. This step allowed some of the best performers—moss phlox, species tulips, tiger lilies, daffodils, daylilies, forget-me-nots, and perennial alyssum—to reseed and multiply on their own.

The Levitans cut paths through the property to connect several "secret" flower beds tucked among outcrops of large boulders. The paths were planted with grass for its soothing effect. An important addition to the garden

*Free-flowing border plantings of azaleas, grapehyacinth, and tulips all bloom together in May.*

was a free-form pool with waterlilies, goldfish, and a cherub at one end who pours water from an urn (with the help of a simple recirculating pump from a hardware store). The pool was backed with a thicket of flowering shrubs, including azalea, forsythia, and redbud; its margins were planted with forget-me-nots, from which huge heads of peony-flowered tulips emerge. Siberian iris extends the garden's early color into late spring.

A number of other ornaments were placed around the garden as focal points: A pair of resting ponies decorate a perennial bed, a rabbit peers out from a rock garden, and a dignified robed maiden stands in a clearing. Several ornate benches provide places to pause and take in a view. The Levitans' favorite source of ornaments is a local wrecking company that specializes in items salvaged from demolished properties.

In midsummer, when the other flowers have faded, Mrs. Levitan's favorite spot is the Tranquility Garden, a series of flower squares bordered by lawn. Each square is bright with annuals—nothing exotic, just marigolds, zinnias, and snapdragons. Bees and butterflies dance everywhere. Flowering shrubs such as rose-of-Sharon and hydrangea provide a colorful background and a sense of enclosure.

The accompanying photographs are testimony to Ruth Levitan's "big bang for the buck" philosophy. They were taken in early May, at the height of spring flowering.

*A red tulip makes a vivid solo appearance in a bed that includes blue forget-me-nots and yellow basket-of-gold. All are perennials that come back year after year.*

# Ensuring Success

*Giving plants what they need is the key to a bountiful flower garden. Whether you buy plants ready grown or start from seeds and bulbs, here's how to choose healthy stock and keep it thriving.*

Why don't my nasturtiums bloom in the summer? Why do the flowers drop off my azaleas just as they are about to open? Why don't my tulips come up? Even the most experienced flower gardener hits a snag now and then, but a solution is usually not all that elusive. Sometimes a problem has a combination of causes, yet often there is just a single cause, such as lack of sunlight, a late freeze, or poor plant stock.

In the pages that follow, you will learn about the major factors that affect flowering and how to choose the best seeds, bulbs, seedlings, trees, and shrubs. You'll learn how to grow flowers from seed, how to transfer ready-grown plants into the garden, and how to start new plants from cuttings and divisions. You'll also learn how to lay the groundwork for an easy garden and keep it performing at its peak.

*Tulips, hyacinths, daffodils, and other spring-flowering bulbs lend easy splendor to the lawn of a suburban home.*

*Sunlight is the most important influence on flowering. Here, coleus thrives in filtered shade, but neighboring scarlet sage prefers direct sunlight.*

## WHAT MAKES FLOWERS BLOOM

Most flowering plants bloom in order to perpetuate themselves. It is their way of attracting pollinators so that they can set seeds to produce a new generation. Some plants flower with such abandon that it seems as if their only purpose is to fill the world with color. Gardeners sometimes say that such a plant "flowers its head off." Azaleas, roses, and French marigolds all demonstrate this ability.

Yet even the most prolific plants require certain conditions in order to flower freely, or indeed at all. The following factors can seriously affect the performance of flowering plants.

### Sunlight

The most important influence on flowering is sunlight. Most flowering plants require at least six hours of sunlight a day to flower. Even the small group of plants that will tolerate light shade—such as impatiens, begonias, and coleus—will not perform well in a heavily shaded garden. Sunlight is so critical that in some cases a 1-percent increase in light levels produces a 100-percent improvement in flowering. Removing a tree limb or painting a fence white may be all it takes to enable sun-loving plants to flower in your garden.

### Temperature

Another influence on flowering is temperature, particularly nighttime temperature. Many flowering plants will survive high noontime heat provided that they have a cool respite during the night. Snapdragons and nasturtiums are examples. Yet other flowering plants are inhibited by low temperature; zinnias, hardy hibiscus, and other plants from southern climates need warm, sunny weather to flower spectacularly.

When certain flowering shrubs fail to bloom, the problem can sometimes be traced to a severe late frost that entered the bud sheaths and destroyed the flowers, yet spared the leaves. Called bud blast, this is a particular problem in early spring, when premature warming can cause buds to break dormancy, only to be hit by a cold blast before they open.

When hardy perennials and hardy bulbs fail to come back the following year, the cause is generally a period of alternating freezing and thawing. During a thaw, dormant plants may start to grow, making them vulnerable to damage by an unexpected freeze.

### Day Length and Time of Day

The blooming of many flowers is influenced by day length and in some cases by time of

*Top: The dwindling light of autumn prompts the chrysanthemums in this border to bloom. Bottom: This garden of flowers that prefer cool weather features tulips, a purple magnolia tree, blue phlox, yellow alyssum, and white candytuft.*

day. For example, chrysanthemums start blooming as days shorten in autumn. Morning glories close in the afternoon and stop blooming entirely as day length dwindles. The moonflower, a type of morning glory, opens only during the evening and part of the morning.

## Watering

All plants need water. Although many flowering plants tolerate drought, a regular supply of water will usually prolong their flowering, especially the flowering of plants such as impatiens and wax begonias, which enjoy cool soil. Since roots can absorb nutrients only in soluble form, they cannot take up these nutrients without water.

In small gardens, people are often quite content to water on an as-needed basis: When a plant shows signs of wilting, they douse it with a garden hose or sprinkler. However, a plant that shows thirst by wilting is already under stress and may never completely recover from emergency treatment. It is far better to make a habit of watering regularly so that plants never come under moisture stress. A rule of thumb is to give established plants a good overnight soaking at least once a week in the absence of rainfall and preferably twice a week during prolonged drought.

## Soil Fertility

A fertile soil has the proper balance of three essential plant nutrients—nitrogen, phosphorus, and potassium. Flowering plants are generally less greedy in their fertilizer needs than are vegetables, but the application of a granular fertilizer, raked or watered into the soil surface at the start of the season, is good insurance. In general, the faster a plant grows, the more fertilizer it needs. If you're not sure of the fertility of your soil, use moderate amounts of a fertilizer with about twice as much nitrogen as phosphorus and potassium.

## Soil pH

The pH of a soil is a measure of its acidity or alkalinity. Wooded areas often have acid (low-pH) soil; deserts tend to be alkaline (with high pH). Although some flowering plants will fare better in one type of soil than the other, most plants recommended in this book will perform well in a wide range of soils.

You can determine the pH of your soil by asking a neighbor who gardens, by purchasing an inexpensive soil test kit, or by submitting a soil sample to a laboratory recommended by your county agricultural agent. Some garden centers will test your soil for free. Minor adjustments to soil pH can be made by adding lime (to acid soil) or soil sulfur (to alkaline soil). However, if your soil pH is extreme, adjusting it may be difficult; instead you may choose to plant your garden in raised beds filled with topsoil from a nursery.

*Applying fertilizer to the soil before planting will make many flowers bloom more vigorously.*

*Removing faded flowers prolongs the blooming of many plants by preventing seeds from forming.*

### Good Grooming

Setting seed for regeneration can drain a plant's energy. Thus, leaving its flowers to wither away and develop seedpods may shorten the plant's ornamental life. One way to keep plants blooming is to *deadhead,* or remove spent flowers, before seed setting can begin. Although deadheading can be tedious, it is less tiresome if done regularly during the cool time of the day.

Fortunately, plant breeders—in their constant quest for improvement—have created some sterile hybrids, which are unable to set seed. Freed from the chore of reproduction, these plants instead pour all their energy into continuous blooming. An excellent example is triploid hybrid marigold, a cross between French marigold (*Tagetes patula*) and American marigold (*Tagetes erecta*).

### Bulb Quality

When bulb growers look at a tulip display and brag that there isn't a blind bulb in the lot, they mean that every bulb has produced a bloom. When buying flowering bulbs, you should make sure to buy only top-size bulbs—bulbs large enough to allow plants to bloom the first year. Many cut-rate bulbs are too small for this and may bloom only the second year or not at all.

These undersized culls have usually been discarded by the bulb industry and purchased by unscrupulous mail-order merchandisers to sell in volume at low prices. Bulbs can be blind for other reasons as well: Sometimes an insect or disease has destroyed the heart of the bulb, or improper curing has caused the bulb to rot. You can tell whether a bulb is healthy by squeezing it gently. A healthy bulb is solid and firm; a diseased bulb may feel spongy.

### Freedom From Pests and Diseases

A healthy environment is important for attractive flowering plants. For example, such summer-flowering bulbs as dahlias are susceptible to thrips and spider mites, which can deform the flowers and thus reduce the impact of a display. Fungal diseases such as yellows (affecting marigolds and asters) can also deform flowers. The best defense against pests is a well-maintained garden, free from weeds and debris and preferably watered by drip irrigation to retard mildew.

### Cultivated Soil

Most garden-worthy flowering plants need a good depth of loose, crumbly soil to allow roots to penetrate long distances in search of water and nutrients. Fortunately, most flowering plants do not need the deep cultivation that vegetables require. Digging to a depth of 1 foot—about one spade length—should suffice for a flower garden, even one planted with flowering trees and shrubs.

*Flowers in bloom look enticing at a garden center but often don't transplant well. It's usually better to "buy green."*

## CHOOSING THE BEST

Perhaps the best advice for home gardeners is the adage "Quality is remembered long after price is forgotten." All things considered, the cost of seeds or plants is the least significant part of any planting enterprise. It is the time and care needed after planting that represent the largest investment. After you have carefully selected a site, prepared the soil, dug a hole, planted, watered, weeded, fertilized, and contended with pests and diseases, you certainly don't want a cheap transplant or tree to die because of poor quality you failed to see.

You can recognize many faults ahead of time—if you know what to look for. However, there is no way of detecting deficiencies in seeds or in mail-order bulbs and root stock. In such cases you must rely on the seller's reputation. Some reliable mail-order suppliers are listed on page 109.

### Buying Seeds and Bulbs

Seeds can be purchased locally at garden centers and even at the supermarket. A wider selection, however, is available by purchasing early in the season through a mail-order catalog. Unlike seeds, which cannot be judged by sight, flowering bulbs can easily be inspected for quality. In general, the bigger the bulb, the better the bloom. It thus pays to buy only those marked "top size" or "flowering size"

and to avoid the smaller culls. Bulbs that are not firm to the touch, or are shriveled or discolored, may be rotten or infested with pests or disease.

Tulip bulbs have a single growing tip; each bulb produces one flowering stem. In contrast, daffodil bulbs are sometimes sold double- or triple-nosed, with two or three growing tips. Double- and triple-nosed bulbs are a better buy than singles; each bulb will produce more flowers.

### Buying Ready-Grown Annuals and Perennials

The quickest way to establish a flower garden is to buy ready-grown plants. For the modest extra cost, you are spared the effort and uncertainty of raising plants from seed. Indeed, many flowering plants are tedious to grow from seed and their propagation is best left to professionals.

**Buying annuals**  When you purchase ready-grown annuals at a nursery or garden center, avoid plants that are already in bloom. It is better to "buy green." A plant that is not yet flowering will almost always survive transplanting better than one in bloom. Because the plant has not had to endure the shock of transplanting while in bloom, it will quickly catch up to the flowering transplant and will put on a better display.

Don't judge a good transplant by its size. Transplants that are tall and lanky are usually under some kind of stress, such as lack of moisture or sunlight, and their roots are probably pot-bound and too small to sustain the top growth that occurs after transplanting. It is much better to choose plants that are short and compact with many side branches. If any plants show a tendency to grow one main stem, pinch out the tip prior to transplanting. Snapdragons, zinnias, and marigolds, for example, often need pinching to encourage bushy growth.

Feel the soil in the pot. If it is bone-dry, the plant is probably already stressed. The soil of all transplants should be moist, and the leafy top growth should be a healthy deep green.

Most annuals are sold in six-packs, but you can sometimes buy them in shallow containers, called flats, with up to 24 plants. Since most flats lack internal dividers, the roots of

*Beautiful tulips are the product of healthy top-size bulbs. Small or diseased bulbs may produce leaves but no flowers.*

*Perennials, such as yarrow, are often sold in gallon containers at garden centers. Look for compact plants with healthy leaves.*

neighboring plants intermingle. When transplanting, be careful to break as few roots as possible. Spread the roots around the planting hole so that they can take off in all directions, and thoroughly soak the soil.

**Buying perennials**   Some perennials, such as coreopsis and columbine, are sold in six-packs like annuals because they grow quickly from seed. More often, perennial plants are grown from root divisions or cuttings and are sold individually in peat or plastic pots. As with annuals, look for compact, stocky plants with a bushy crown of leaves. Avoid plants with

brown leaves, plants coming into flower, and plants that have weak, gangly stems.

To remove a perennial from its pot, soak the soil and slide out the rootball. If this proves difficult, cut the sides of the pot, which will make it easier to remove the plant without tearing its roots. Tease the roots apart and plant so that the rootball fits snugly into the hole. Firm the surface with a trowel or with the sole of your shoe to make good soil contact around the roots; then water. In the absence of rainfall, newly planted perennials need watering almost every day until they are well established.

## Buying Trees and Shrubs

Most trees and shrubs are sold in containers or "balled and burlapped," with burlap encasing a ball of soil around the roots. In either case, choose plants whose rootball or container has been sunk into a pile of peat or wood chips to keep the roots cool. Trees and shrubs left out in paved display areas often suffer root damage from burning heat.

Inspect the base of the trunk for signs of girdling by mice or deer. These animals love to gnaw bark around the base of a tree, and if they encircle it completely the tree will die after transplanting. Inspect for this damage deliberately; it is easily overlooked.

To remove plants from metal or plastic containers, try soaking the roots with water and pulling gently but firmly to release the plant from the pot. If this fails, you may need to cut away the sides of the container to remove the rootball without damage. Plants that have been in containers too long may be root-bound, with a massive tangle of roots that must be teased apart and spread as wide as possible to help them grow freely in all directions. Some shrubs and trees are sold in peat pots that will gradually decompose when immersed in soil. However, even with these it is best to take a sharp knife and cut away the pot to release the rootball, which may be too tangled to grow efficiently.

You don't need to remove burlap from a rootball. Just loosen the top and cut away string tied to the trunk. The burlap soon rots when it is covered with soil, leaving the roots unhampered.

To plant a tree or shrub, dig a hole almost as deep as the rootball and twice as wide. Sink the rootball into the hole and fill in the sides with topsoil. For a 1-gallon plant, sprinkle about 1 tablespoon of a concentrated fertilizer into the soil as you fill the planting hole. If you are using a less concentrated fertilizer, such as steer manure, add about ½ cup. The top of the rootball should be even with, or slightly higher than, the soil surface. Create a catch basin for water by scraping soil into a ridge around the plant.

Water thoroughly around the base of the tree or shrub and cover the soil with a layer of wood chips, shredded leaves, or similar organic material to retain moisture and keep weeds down.

**Planting From Containers**

Soil from planting hole

Twice rootball width

Same as or slightly less than rootball depth

**Planting Bare-Root Stock**

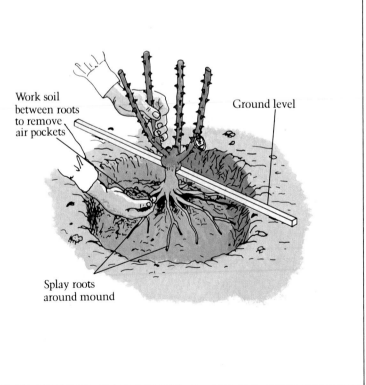

Work soil between roots to remove air pockets

Ground level

Splay roots around mound

## Finishing Up

Base of plant at or above basin level

Watering basin

## Buying by Mail

Mail-order companies offer an enormous selection compared to that of garden centers or nurseries. Seeds and bulbs, in particular, are easy to purchase through catalogs because they are lightweight and sustain little or no damage in transit.

Live plants, however, are a little riskier. They are normally shipped dormant—with perhaps a few bare branches showing above the roots, or a single stem in the case of a tree. To keep freight costs down, most are shipped as "bare-root" stock, with soil washed away and roots cushioned in a wad of damp sphagnum moss, shredded newspaper, or sawdust. Shipping plants this way can take its toll, so inspect plants carefully when they arrive. Superficial damage such as a broken root or snapped-off branch tip can be remedied by pruning away the broken piece. However, if the root mass shows signs of rot or fungus growth, or if the main stem is broken in two, request a replacement or refund.

Bare-root stock should be put into the ground as soon as possible after it arrives. Before planting, immerse the roots in a bucket of water so that they can take a long drink.

When planting, make a mound of soil at the bottom of the hole and splay the roots around it in an octopus formation.

In the world of mail-order merchandising, there are general practitioners and specialists. If you want to make a collection of a particular plant genus—such as daylilies, iris, or daffodils—seek out companies that specialize in that plant group. Many of these firms offer special varieties available nowhere else and have developed custom shipping containers to ensure safe passage of their products. Also, their plant stock will usually be superior to anything in a general catalog.

Some catalogs may offer a choice of sizes. For example, a rose catalog may advertise #1 grade, #1½ grade, and #2 grade plants at progressively lower prices. The lower-priced plants are usually the runts or culls and should be avoided.

A final caution: The mail-order world is full of get-rich-quick schemes. If it sounds too good to be true, it probably is. It is wise to buy from well-known companies whose catalogs avoid puffery. The companies listed on page 109 of this book are generally reliable and offer guarantees.

*Black-eyed-susan
blooms reliably from
seed sown directly
into the garden—even
on top of frost or
snow—making a
display that comes
back year after year.*

## STARTING PLANTS

There are many reasons why you may choose to start plants yourself. Sometimes the local nursery may not have the ready-grown variety you want. Sometimes, as with California-poppies, a plant cannot survive transplanting and must be sown in the garden from seed. Perhaps the largest incentive of all is the money you can save. For less than the price of a six-pack of annuals, you can sow a packet of seeds that will yield a 15-foot row of flowers. Moreover, some of the best plantings are free, in the form of cuttings from your own or other people's gardens.

Although more time-consuming than buying ready-grown plants, growing seeds, bulbs, and cuttings can be quite simple. The instructions that follow will help you get started.

### Starting Plants From Seed

Seeds are infinitely variable—not only in size and form but in ease of germination. Some annuals, such as sunflowers, have large, easy-to-handle seeds that germinate fast and produce plants that flower quickly. Other seeds are so tiny, and need so much fussing over, that it's better to leave their propagation to

professionals. Wax begonias, impatiens, and petunias belong to this category.

Between these two extremes is a vast range of reasonably easy-to-grow seeds. These can be sown directly into the garden or started indoors if earlier flowering is desired. For example, marigolds, asters, and straw-flowers sown directly into the garden will bloom by midsummer, but if they are started indoors six weeks before the outdoor planting dates, they will flower several weeks earlier.

**Direct seeding**   Sowing seed directly into the garden is the easiest way to start seeds, but not all annuals and perennials can be reliably started this way. Seeds best for direct sowing are usually easy to handle, fast germinating, and fast growing; for example, those of cosmos, marigolds, and zinnias. Some very small seeds, such as those of Shirley poppies, are sown directly because they do not transplant well.

There are two kinds of direct seeding—in straight rows (or furrows) and in groups. Straight-row seeding is popular for cutting gardens; group-seeding is useful for large beds and wide borders where an informal, patchwork effect is desired. Regardless of the method you choose, the soil surface must first be raked to a fine granular texture.

To sow seed in straight rows, stretch a line and make a shallow furrow with a trowel to the depth recommended on the seed packet. Sow seeds in the furrow at the required spacing and cover them with fine soil. Ensure good contact between seeds and soil by gently pressing the soil with your foot or with a board. Water immediately and keep the soil moist until the seedlings appear. If the germination rate is high and the seedlings are very close together, thin them by pulling up the weaker plants. The optimum distance between thinned plants is usually specified on the seed packet.

To sow seed in groups, take a stick and etch into the soil an outline for each sowing area. Scatter the seed over the surface and either press the seeds into the soil with a board or rake the seeds in, covering them with a light sprinkling of soil or sand just sufficient to anchor them. Water and keep the soil moist until the seeds have germinated. You may also need to thin the seedlings to prevent overcrowding.

**Starting plants in pots** Starting plants indoors can give them a jump on cold weather, allowing them to flower earlier or to survive in climates colder than what they are used to. You can start seeds in peat pellets, peat pots, or plastic containers—the last is least expensive but somewhat less convenient.

Peat pellets are hard, round disks of compressed peat. When placed in a tray of water they expand to several times their original size and become soft enough to allow you to press seeds into a cavity on the top. Some peat pellets are held together with a chemical binder that allows roots to grow through it unrestricted; at transplanting time you simply plant pellet and all. Other types of peat pellets are held together with an expandable netting. Roots can grow through this netting, but you should gently remove it (netting peels away easily) so that roots can grow freely.

Peat pots are compressed-peat vessels that may also be buried directly into the ground, but they must be filled with potting soil in which seeds can grow. Use a commercial soil mix that has been sterilized to kill diseases and weed seeds. Whatever type of peat container you use, remember that it loses moisture from all sides and may therefore need frequent watering.

Plastic pots are less expensive than peat containers, but you must remove seedlings from them before transplanting. If you use plastic containers, make sure they have drainage holes; otherwise, plants may become waterlogged.

To plant in peat or plastic containers, place two or three seeds on top of the peat or soil in each container, and cover the containers with clear plastic to keep them from drying out. Keep the containers warm (70° to 80° F) by setting them on top of a TV, refrigerator, or water heater. Avoid direct sunlight, which can overheat the seeds. When the first seedlings appear, remove the plastic and pinch out all but the strongest plant in each pot. Then move the containers to a bright window. Sun-loving plants need four to six hours of direct sunlight daily; shade plants need lesser amounts of indirect or filtered sun. Once started, young plants thrive at temperatures between 68° and 72° F.

When seedlings are about 4 inches high, with thick stems and many leaves, transplant

them into the garden. If you've used peat containers, simply bury them in a hole deep enough to cover the top of the container with about ½ inch of soil. Make sure that the rims of peat containers are completely covered; if they protrude above the soil they can act as wicks that will dehydrate the plants.

If you've started plants in plastic containers, slide a knife along the side of the container to ease each plant out of its pot, keeping the rootball intact; plant flush with garden soil. Once new plants are in the ground, water them immediately.

Seeds that are very small, and thus are hard to handle individually, may be easier to start by the two-step method—so called because seedlings are transplanted twice. First, sow the seeds together in a large, flat container filled with potting soil. When seedlings come up and their second set of leaves appears, use a sharp pencil to pry each seedling out. Carefully transfer the young seedlings to individual pots and transplant them into the garden when they are about 4 inches high.

*Plants started in peat pots can be transferred directly into the soil. Be sure to cover the rim of the pot—otherwise it acts as a wick that can dry out the plant.*

## Starting Bulbs

Bulbs, like plants grown from seed, are classified as hardy or tender. Hardy bulbs, such as daffodils and tulips, can withstand mild frosts and freezing; tender bulbs, such as dahlia and lily-of-the-Nile, are killed by freezing. The cold hardiness of particular bulb varieties is identified in catalog descriptions and on bulb-package labels.

A good selection of easy-to-grow bulbs can be found in garden centers and even hardware stores and supermarkets. However, a far larger selection of bulbs is available through mail-order catalogs specializing in them. Spring-flowering bulbs, such as hardy tulips and daffodils, are offered for sale after September 1 for planting any time during late summer and early fall. Provided the ground stays unfrozen, they can be planted up until Christmas. Summer-flowering bulbs, such as gladiolus and dahlia, normally go on sale at Eastertime for planting outdoors until the end of May.

Most bulb losses are caused by rotting, due to poor drainage, and by hungry rodents. Tulips and crocuses are especially vulnerable to rodents. To deter rodents, cover plantings with a length of fine wire mesh that has a gauge wide enough to permit leaves and flower stems to grow through, yet small enough to prevent squirrels and mice from digging up the bulbs. Rodent-repellent flakes, or mothballs buried with bulbs, also help to repel rodents. Where gophers are a problem, dig a larger-than-usual hole for each bulb and line it with 1-inch-mesh chicken wire to make an underground barrier.

Deer are fond of tulips, so you should plant these bulbs where deer are unable to forage for them. Alternatively, spray the leaves with a deer repellent as soon as they strike through the soil in early spring. Deer repellents are odorless sprays that make plants taste unpleasant. Garden centers and mail-order catalogs usually stock these repellents.

When planting flowering bulbs, pay particular attention to the recommended planting depth, and take the measurement from the base of the bulb, rather than from the tip. Some bulbs, such as gladiolus and dahlia, will not survive winter in northern states and must be planted anew each year; alternatively, the

*Place bulbs pointed-end-up and use a trowel or bulb planter to bury them at the recommended depth, as measured from the bottom of the bulb.*

bulbs can be lifted after the tops have died down in autumn and stored in a frost-free area indoors for replanting in spring. To keep spring-flowering bulbs coming back each season, it's essential to allow the leaves to die back naturally, since the leaves help produce the food needed to support growth the following year.

## Multiplying Plants by Division, Layering, and Cuttings

As an alternative to buying seeds or transplants, you can easily obtain new plants through division, layering, and cuttings. These asexual methods of reproduction produce a plant that is identical to its parent.

Most perennials are easily divided, since they spread naturally by underground stems or roots. Any section of root with a growing point (underground bud) can be separated from the main plant and replanted to start a new one. Daylilies, Shasta daisies, and Siberian irises are a few of the many perennials that can be readily divided.

Many annuals, perennials, shrubs, and trees are easily propagated by cuttings. Soft-stemmed flowering annuals, such as coleus and impatiens, can be propagated within two weeks from 4- to 6-inch sections of stem placed in water. Once roots have formed, the plant can be transferred to the garden.

Some perennials, including chrysanthemums and sedums, can be rooted easily from 6-inch tip cuttings placed in moist, sandy potting soil. Cuttings from forsythia and some other shrubs will also take root in moist potting soil, especially if the cut end is first dipped in a rooting hormone. This is a powdery substance that can be obtained from garden centers.

A number of desirable shrubs, such as azaleas and roses, increase readily by layering. Simply grasp a long, low branch, peg it to the soil with a piece of bent wire, and scrape away the bark where the stem touches the soil. Within six months to a year, this section of stem will take root at the soil line; it can then be cut free and transplanted.

*In layering, branches are pegged to the soil and cut free when they take root, usually within six months to a year.*

# 15 Easy Steps to Success

Creating and maintaining a beautiful flower garden can be easy if you follow the simple procedures that follow.

1. Make your plans on paper before you set foot in the garden. Plot the size and content of your garden using graph paper, with each square of grid representing 1 square foot of garden space. If your garden is too large or complex for a single sheet, use separate sheets for different beds and borders. See page 27 for instructions on laying out these areas, and see the fourth chapter for more general tips on garden design.

2. Test the soil to determine its pH and nutritional status, or ask a gardening neighbor about conditions in the area. Add lime to neutralize acid soil, sulfur or peat to neutralize alkaline soil, and other amendments such as compost to improve the soil structure. Evaluate drainage, and avoid planting in low spots or other problem areas. See page 29 for a discussion of soil preparation and the third chapter for help with problem soils.

3. Determine how much sunlight your garden receives, and let this guide your choice of plants. Most flowering plants require at least six hours of sun each day, but this need not be direct sunlight. See page 42 for tips on adding light to shady gardens, and page 62 for a list of plants that thrive in shade.

4. Lay out new beds and borders with string stretched between stakes; use flexible garden hose to define curving edges. See page 27 for garden layout guidelines.

5. Condition soil that contains too much sand or clay by applying peat moss, compost, or other soil amendments and working them into the soil with a garden fork. See page 29 for advice on preparing the soil.

6. Dig the planting site to a depth of at least 12 inches for most bedding plants or to the depth of the rootball for trees and shrubs. Remove large stones and extraneous roots. Break up clods with the edge of a spade. See page 20 for tips on digging holes for trees and shrubs.

7. Apply a fertilizer that is suitable for flowers and, if possible, tailored to the deficiencies of your soil. (Soil evaluation by a soil-test lab will tell you what formula to use.) See pages 30 and 31 for information on fertilizing.

8. Rake the site level. Use the front and back of the rake to make a fine, crumbly surface.

9. Plant seeds or transfer ready-grown plants from containers. Plant cold-hardy varieties several weeks before the last frost date; plant tender varieties after the last expected frost date. See page 18 for tips on buying seedlings; refer to page 22 for seed-starting instructions.

10. Water seeds or transplants thoroughly at the time of planting. In the absence of rainfall, water daily until established and weekly thereafter. Even a week without rainfall can cause moisture stress in established plants and slow down growth. See page 34 for a discussion of watering.

11. Thin seedlings, leaving room for the strongest plants to develop to maturity. Most flowering annuals need to be spaced 6 to 12 inches apart; perennials may need more room. Use scissors to decapitate unwanted seedlings instead of pulling up roots, which can disturb neighboring plants.

12. Weed regularly and place mulch around plants to help control the weeds. Organic mulches such as cocoa-bean hulls, shredded pine bark, and wood chips are the most attractive types for flower gardens. See pages 31 to 33 for mulching guidelines.

13. Control insects, pests, and diseases; check with experienced gardeners to learn which of these are most troublesome in the area and what controls are most effective. Flower gardens are less vulnerable than vegetable gardens, but slugs, snails, rodents, or deer may need special controls. See page 41 for details.

14. Remove faded flowers from long-blooming plants to prevent seed formation, which can steal energy from flowering. See page 17 for information.

15. Clean up the flower garden in fall. Remove annuals and cut down dead flower stems of perennials, which can harbor harmful organisms over the winter. Divide any overgrown clumps of perennials; see page 25 for instructions.

# PLANNING THE GARDEN

Many gardeners boast that they don't plan their gardens—they just scatter a few seeds here and there in a rectangle of soil, and stick in a few plants purchased on impulse from a local garden center. Although the result of this spontaneity may look fabulous, formal planning will save you time and money and will produce a more dependable garden.

Proper planning begins in January with the arrival of mail-order catalogs. Use these to make lists of plants you wish to grow and to help you decide which to order by mail and which to buy locally.

Next comes sketching the garden on paper to help you decide where to place everything on your wish list. Ideally, you should design the garden before ordering or buying plants. Sketching the flower areas will help you determine how many plants or packets of seeds to buy and whether your garden can supply the sun or shade they need.

The easiest way to plan a garden layout is with graph paper. Use one sheet to sketch the property, showing where beds and borders will be located and marking whether they are in sun or shade. Then use a separate sheet to plan each bed, allotting one square of grid for each square foot of garden space.

### Estimating Quantity

Most annuals are happy with 12-inch spacing, but perennials may need as much as 3 feet. This rule of thumb can help you determine how many plants you will need for a given area. Shade in the beds and borders you have sketched on graph paper with felt-tipped colored pens. Use different colors to represent different plant groups; for example, 12 yellow squares to show a planting of a dozen marigolds or 24 red squares to show a planting of two dozen wax begonias. Then draw up your plant list. To create a visually interesting bed or border, remember to include low edging plants and tall background plants.

### Laying Out Beds and Borders

Beds and borders are the most common locations for annuals and perennials. A bed usually is an island of soil—either square, rectangular, oval, round, or kidney-shaped—surrounded by paving or lawn. The most pleasing way to plant beds is to group tall

plants in the middle, then circle them with intermediate-height plants and an outer ring of low-growing plants.

A border normally is a strip of soil backed by a hedge, wall, fence, or path. The planting may be formal (neatly geometrical) or informal (with a wavy edge). Tall plants should be placed at the rear, intermediate-height plants in front of them, and low-growing plants as an edging. A line of shrubs such as forsythias or hydrangeas makes an excellent background for a border of shorter flowers. See pages 49 to 50 for a further discussion of beds and borders as design elements.

To transfer your planting plan from a piece of graph paper to the garden, take a sharp stick or a bag of flour and outline the planting stations for each flower variety. For greater precision, mark out beds and borders with string stretched between wooden stakes, or use flexible garden hose for a curving edge.

*When planning a garden, try to visualize how plants will look when fully grown and in bloom. Top: Pinkish hues predominate in this stand of azalea, dogwood, and rhododendron. Bottom: Bench-height daffodils and 'Darwin' hybrid tulips help frame a sitting area.*

*'Darwin' hybrid tulips are planted in fall to come up shortly after daffodils in spring.*

### Scheduling Planting

Before you start to plant, first determine which plants are frost tolerant and which are frost tender. For example, pansies, calendulas, and snapdragons will survive light frosts and can be planted out several weeks before the last frost date in your area. Begonias, petunias, and impatiens, however, are highly susceptible to frost damage and should not be placed outdoors until frost danger has passed.

Spring is usually the time for planting, whether you choose herbaceous plants (annuals, perennials, and bulbs) or woody plants (trees and shrubs). Consider when you want the plants to bloom; you can regulate the flowering of many plants by planting them late. For example, if you plant dahlias at the end of June, they will not flower until early fall, which will give the garden color right up to the first fall frost.

*Compost of decayed leaves or animal manure adds nutrients to depleted soil and neutralizes acidity or alkalinity. Spread compost on top of soil, then rake it in to a depth of 3 to 4 inches.*

## PREPARING THE SOIL

Before setting plants into the ground, it pays to get the soil into tip-top condition. Without good soil, achieving a fine flower garden will be an uphill battle. Soil not only anchors plants but holds the moisture and nutrients they need to grow and flower. Soil that is too sandy lets these drain through too quickly; soil containing too much clay is hard for water and plant roots to penetrate. The ideal soil, called loam, is crumbly and well aerated, with plenty of humus (organic) content. In a loamy soil, plant roots can travel long distances in pursuit of nourishment to support vigorous leaves and masses of colorful flowers.

To make your soil more loamy, you must load it with plenty of humus—organic material in the form of compost, well-decomposed animal manure, leaf mold, or bales of peat. Even soils that have been well maintained need to be topped up with organic material every year because the natural workings of sun, wind, and rain tend to leach out nutrients and pack down the soil. Soil conditioner will also help neutralize overly acid or alkaline soil, which can impede the growth of some flowering plants.

Of the soil conditioners most commonly used, leaf mold is the most valuable because it holds moisture extremely well and because it has an ample supply of nutrients. Peat is the least desirable because it lacks nutrients. Buy peat or manure at a garden center, or check classified ads in a newspaper for sources of less-expensive compost. Leaf mold is so sought-after as a soil conditioner that many nurseries hoard it for their own use and rarely sell it. You can make your own by piling leaves into a corner of the garden and letting them rot down into lightweight, fluffy, rich humus. To speed this process, shred the leaves first with a lawn mower or leaf shredder. If you use a lawn mower, simply run the blades over small piles of leaves. Spade the soil conditioner over the soil surface to a depth of 3 to 4 inches and rake it in.

Cultivate the soil to the depth of a garden spade (8 to 12 inches), and comb the soil surface with the prongs of a garden rake. Use the flat back of the rake to break up any clods and make the surface even finer. When applying fertilizer, sprinkle granules over the soil surface before raking, at the rate recommended on the package. Take care not to tread on newly dug soil. If you must do so in order to plant, put down a wooden board to cushion your weight.

## MAINTENANCE TIPS

Once you have planted your flower garden, there are three essential maintenance routines—fertilizing, weeding, and watering—that will keep it at its best. These routines may sound like a lot of work, but they needn't be. One application of fertilizer at the start of the season will usually suffice, weeds can be controlled by mulching, and water can be applied effortlessly by means of drip hoses or lawn sprinklers. You'll find that if you set up a regular schedule for garden maintenance and use the timesaving techniques outlined on the following pages, you'll need to spend no more than a few minutes each day maintaining your garden.

### Fertilizing

To remain healthy, every plant needs a balance of three essential nutrients—nitrogen, phosphorus, and potassium (also called potash)—that are present to varying degrees in most soils. Of the three, nitrogen is the most quickly depleted by plants or leached away by rain; phosphorus and potassium also dissolve away, but more slowly. To make up for deficiencies, all three need to be supplemented with natural or commercial fertilizer.

Natural fertilizers include leaf mold, blood meal, decomposed manure, bonemeal, and garden compost. Commercial fertilizers are synthetic mixtures sold in the form of liquids or granules, in various strengths and formulations. If you want to do only a minimum of fertilizing—such as a token application before spring planting—you should grow only annual flowers.

To keep perennials healthy, it is good practice to fertilize them twice a year—once in spring before dormant roots begin to grow, and again in fall after the first frost. (Warm-climate perennials need fall fertilizing only.) The easiest way to fertilize is to sprinkle dry granular fertilizer around the bases of plants and water it into the soil surface. Alternatively, you can apply liquid fertilizer to the soil in a dilute solution (follow instructions on the product label). Liquid fertilizers are generally faster acting than granular fertilizers and more economical.

*A healthy garden has soil that provides plants with essential nitrogen, phosphorus, and potassium. Annuals need fertilizing once a year, perennials twice.*

To prevent plant roots from being burned, use no more fertilizer than is recommended on the label. It's also best to wait 10 days before setting out plants in newly fertilized beds and borders, to allow the fertilizer time to permeate the soil.

Plants can also absorb nutrients through their leaves, making it possible to feed them this way rather than by applying fertilizer to the root zones. Called foliar feeding, this technique is useful when plants are growing in sandy or gravelly soil, which does not retain nutrients well. To foliar-feed plants, dilute liquid fertilizer with water according to directions on the label and spray the solution onto both sides of the leaves. Tests have proven that liquid fertilizers with a 1-2-1 ratio of nitrogen, phosphorus, and potassium (such as 5-10-5 or 10-20-10 fertilizers) are especially effective for foliar feeding.

Understanding how fertilizers are formulated will help you to purchase them more confidently. When you buy a bag labeled "10-15-10," you are getting a mixture that contains 10 percent nitrogen, 15 percent phosphorus, and 10 percent potassium—a total of 35 percent active ingredients. The other 65 percent is filler, a distributing agent that helps the fertilizer to be applied evenly. Thus, a 10-pound bag of 10-20-10 fertilizer contains twice as much nutrient material as a 10-pound bag of 5-10-5 fertilizer. If both bags cost the same, the first is obviously the better buy.

High-nitrogen formulations such as 10-5-5 are particularly suitable for shrubs and trees. Because nitrogen is quickly used up by plants and easily leached from the soil by rain, some fertilizers offer nitrogen in a slow-release form that remains in the soil longer. Slow-release fertilizers are usually worth the extra expense. High-phosphorus formulations such as 5-10-5 are best for flowering plants, particularly annuals, perennials, and flowering bulbs. Bonemeal is a natural source of phosphorus, which makes it a good choice if a soil test shows mainly a deficiency of this element. Phosphorus is normally slow acting and remains in the soil much longer than does nitrogen (unless you have chosen a fertilizer whose nitrogen is the slow-release type).

There are other nutrients, called trace elements, that plants need for health and vigor. Boron, calcium, and iron are a few. Although

many commercial fertilizers include them, they are required in such small amounts that good soil management—such as the addition of compost—generally provides them in sufficient quantity.

## Mulching

A mulch is any ground covering that serves as a buffer between the soil and the air. There are many types of mulches—including black plastic, tree bark, and gravel—but all serve the same purposes: to smother weed growth, retard water loss, regulate soil temperature, and reduce erosion. The type of mulch you choose will depend on your budget and aesthetic preferences, and on the selection available in your area. You'll also need to weigh the

*In a technique called foliar feeding, diluted liquid fertilizer is siphoned through a hose and sprayed directly onto leaves. This is useful where soil is too sandy or gravelly to hold nutrients.*

advantages and disadvantages of the various types of mulches, as outlined below. No matter what type you choose, mulching is one of the best labor-saving techniques you can use.

Mulch materials can be classified as natural (straw, leaf mold, pine bark, hay, and sawdust, for example) or artificial (black plastic and woven soil blankets, for example). In the eastern and midwestern United States, many garden centers offer wood chips, bark chips, shredded pine bark, and cocoa-bean hulls as natural mulches. On the West Coast and in the Southwest, cedar chips, fir-bark chunks, peat moss, and redwood-bark chips are more commonly available. In the South, shredded fir bark, pine chips, and cottonseed hulls are common. Inexpensive black plastic, available everywhere, is by far the most popular artificial mulch.

Natural mulches give a tidy appearance to flower beds and borders. They also keep the soil from forming a crust that prevents air and moisture from penetrating to the roots. Good aeration encourages earthworms, which enrich soil with their castings. Some mulches, including leaf mold, add nutrients to the soil as they decompose. Others, such as cedar and redwood chips, repel insects from plants.

Unfortunately, there are certain disadvantages to most mulches. Sawdust and shredded bark can deplete the soil of nitrogen as they decompose, thus working against any fertilizer you may have applied. Straw can harbor rodents, slugs, and fungal diseases, and may carry weed seeds, which will actually increase your weed problem. In dry areas, straw and peat moss may be a fire hazard. Black plastic and other artificial mulches can overheat the soil and promote mildew growth by trapping moisture; they also look unattractive. Finally, the better-looking natural mulches can be expensive. You should weigh these disadvantages against the advantages when choosing a mulch.

**Using natural mulches** Natural mulches shade the soil surface and help retain moisture, thereby lowering the soil temperature. They should thus be applied to the garden only after the soil has had a chance to warm up in spring. In summer, however, these same properties provide plants with a welcome respite from the heat.

Apply at least 2 inches of natural mulch to every bed or border. Keep a space clear around the plant stems to prevent rot and disease—particularly with newly transplanted annuals, perennials, and roses. Mulch seedbeds only after the seedlings are up and well established.

To help hardy perennials and roses survive the winter, apply natural mulches after the ground freezes. A layer of mulch will help to keep the ground frozen during those brief warm spells that can fool plants into breaking dormancy too soon.

Beds and borders mulched with natural material look most attractive when they have well-defined boundaries, such as a sharply edged lawn or a raised border of bricks. These edgings also help to keep the mulch inside the bed or border.

**Using artificial mulches** Black plastic is a popular artificial mulch because it is inexpensive and easy to apply. Laid over the soil early, before planting, it will raise the soil temperature and give plants a head start.

The main disadvantages of black plastic are its unsightly appearance and its impermeability; rain does not soak through it, and soil that is too wet cannot dry out. You can overcome the unsightly appearance of plastic by covering it with a light layer of shredded leaves, peat moss, or similarly attractive natural mulch, which will also keep the soil from overheating. A similar plastic covering called a mulch blanket comes punched with holes that aid evaporation; some mulch blankets are brown to match the soil.

Artificial mulches are particularly useful in areas where plants are growing in rows,

such as in a cutting garden or a rectangular border. They are also ideal for covering raised planting beds because they match the beds' rectangular form. If you make the beds about 2 feet wide and heap the soil about 4 to 6 inches high, you can cover them efficiently with 3-foot-wide rolls of black plastic. After laying down the plastic, anchor the edges with soil. Cut holes at 12-inch intervals along each planting row to mark stations for seeds or transplants.

Growing plants through plastic is especially effective if you install drip irrigation hoses (see next section) underneath the plastic. The layer of artificial mulch slows evaporation and protects the hoses from damage by stray animals. With hoses in place, the entire area can be watered at the turn of a faucet.

*Left: A natural mulch of wood chips controls weeds around this planting of petunias, impatiens, and marigolds.*
*Right: Newly planted French marigolds poke through holes cut in black plastic, an inexpensive artificial mulch.*

## Watering

Without water on a regular basis, even the most drought-resistant plants will die. Just a week without water will put many flowering plants under stress, because water not only helps keep the soil cool but dissolves nutrients so that the plant roots can absorb them.

For some small gardens, such as those made up of container plantings, an inexpensive watering can is sufficient. For larger spaces of up to 250 square feet, a garden hose with an adjustable nozzle—or better yet, a watering wand that attaches to the end of the hose—can be used. Larger beds and borders are watered more efficiently by turning on a lawn sprinkler in the early evening, after the sun's burning rays have diminished. However, this kind of overhead watering should be avoided if possible—especially in hot, muggy weather—since wet foliage encourages fungal diseases such as powdery mildew. If this is the only easy way to water, however, it is better than not watering.

**Using drip irrigation systems**   Perhaps the most valuable investment a gardener can make is a drip irrigation system—a network of hoses that emit a trickle of water all along their length. At the twist of a faucet, this ingeniously simple and inexpensive system sends

*Top: Plants in porous pots lose moisture rapidly and may need frequent watering. A watering wand attached to a hose is ideal for reaching into awkward places. Bottom: Plastic micropore hose is useful for larger gardens. Water seeps through its tiny pores, watering plants efficiently at the root zone.*

*All but the most drought-tolerant plants need weekly watering in spring, twice-weekly watering in the heat of summer. Avoid using sprinklers, which promote mildew.*

water directly to each plant, delivering it slowly and steadily for maximum absorption. Unlike a lawn sprinkler, a drip system uses water efficiently, applying it only where it is needed. And since the leaves remain dry, plants are less susceptible to mildew. Moreover, a drip system stays in one place, freeing you from the chore of hauling heavy cans or hoses around the yard.

Drip systems for a typical 500-square-foot garden cost about $30. One type, called an emitter system, has tiny emitters, or nozzles, spaced at intervals of about 2 feet. Emitter systems are relatively expensive and tend to clog. Cheaper and better are the "micropore" hoses made of special plastic or rubber that "sweats" moisture all over through millions of tiny pores. One white plastic micropore system is so reasonably priced that you can discard it after one season, though it may last for up to three seasons if you store it under cover during winter. A black type, made from recycled rubber tires, is several times more expensive than the white kind but it will last indefinitely.

**Knowing how and when to water** No irrigation system can improve on a steady overnight rain—although drip irrigation comes close. When watering plants you should try to simulate natural rainfall. Standing around for a few minutes with a garden hose does very

little for plants because the water cannot penetrate deeply into the soil. Similarly, watering large areas with a can is not only backbreaking but often fails to distribute water evenly.

In the absence of rainfall, all but the most drought-tolerant plants need weekly watering in spring and twice-weekly watering during the heat of summer. Watering in the late afternoon or early evening is best, because it minimizes water loss through evaporation. Obviously, trees and shrubs—particularly large specimens—will need more water than annuals. Container plantings also need frequent watering.

Never wait for plants to wilt before you water them. Wilting is a signal of distress; plants may never recover even if watered promptly. To catch plants before they wilt, make a habit of feeling the soil at the base of the plants. Grab a handful of soil and crumble it. If it is dust-dry, water immediately. If soil particles feel damp and stick to your fingertips, there is still moisture available. Avoid overwatering. Too much water can induce root rot and fungal diseases.

To prevent rapid moisture loss from your flower beds or borders, apply 2 to 3 inches of a natural mulch such as peat moss or pine bark to the soil around the plants. Mulch trees and shrubs as far out as the dripline (the limit of the branches).

# Overcoming Hurdles

*It's easy to give up when things go wrong in the garden. Here are some typical problems that flower gardeners face and some helpful solutions.*

The perfect planting site rarely exists. What we all desire is crumbly, well-drained soil on level ground, in sun or light shade, sheltered from fierce winds and watered by frequent, gentle rains that fall only during the night. A moderately long growing season and an absence of pests and diseases would be sublime. Unfortunately, few gardens possess anything close to this idyllic combination.

Featured here, in alphabetical sequence, are some of the most troublesome obstacles to easy flower gardening. Don't be intimidated by the list; even a site that seems to have everything going against it can be modified in some way to allow an assortment of flowers to grow there. Consider the determination of Napoleon: One of his last feats was to grow flowers on the harsh, inhospitable isle of St. Helena in the south Atlantic, where he was exiled by the British. To relieve boredom and maintain morale among his followers, Napoleon set to work building stone walls and wattle fences to shelter a space from the relentless wind, and hauling compost to make raised beds atop the impervious rock. More than one hundred and fifty years after Napoleon's death, this garden still thrives as a testimony to his persistence.

*A shady garden is inhospitable to many flowering plants, but ideal for red coleus, pink wax begonia, and broad-leaved hosta. These thrive in cool, moist conditions.*

*City gardeners
must often cope with
small planting areas
and extreme light
or shade. Here,
marigolds and zinnias
thrive in a walled
urban garden.*

## CITY SURROUNDINGS

City homes are often surrounded by asphalt, concrete, and other paved surfaces, with not a scrap of soil available for gardening. Rooftops, too, often have flat areas that beg for color and greenery. The answer, of course, is to make a container garden. A number of flowering annuals, perennials, bulbs, shrubs, and trees will grow readily in containers. A list of easy-care varieties for container planting appears on page 58.

Even a windowsill, balcony, or flight of outdoor steps is sufficient space for several potted plants. Walls can be hung with baskets or window boxes mounted on metal brackets. A long-handled watering wand that fits the end of a hose will help you to reach inaccessible planters.

On rooftops and balconies, be careful to place planters so that they cannot fall on pedestrians. Also make sure that the structure is strong enough to carry the extra weight.

City gardens often suffer from extreme light or extreme shade. A cool, shady location in the shadow of a building will usually accommodate shade-loving plants such as impatiens and coleus, but an exposed site may be oppressively hot even for plants that love the sun. Brick and stucco walls reflect light and heat, and can literally burn the life out of plants. If this is a problem in your garden, consider covering the walls with a leafy vine such as Virginia creeper (*Parthenocissus quinquefolia*) or English ivy (*Hedera helix*), or cover the wall with a dark-colored lattice to absorb some of the sun's rays. In sunny areas, use wood and clay containers instead of plastic or metal ones; the evaporation from porous pots keeps roots cool. If you must use plastic or metal, choose containers with a double-wall construction; the space between the walls acts as insulation.

## CLAY SOIL

Many areas have clay soil, composed of particles that pack into a sticky mass when pressed together. Water usually puddles on the surface for a long time, unable to penetrate the soil efficiently. Roots have difficulty growing through clay soil, and plants become stunted.

To convert clay soil into good garden loam, you need to load it up with plenty of humus in the form of compost, well-decomposed animal manure, leaf mold, or bales of peat, working the humus into the surface with a garden fork. Since clay is heavy to work with, it's less labor-intensive to build raised beds with landscape ties, which resemble railroad ties and are available from garden centers and lumber dealers. Within each frame of landscape ties, lay a 4- to 6-inch base of crushed stones over the soil as a drainage field. Then pile on the humus. Soil in a finished bed should be at least 1 foot deep, giving roots plenty of room to grow.

## COASTAL WIND, SAND, AND SPRAY

Perhaps no sight is lovelier than a drift of flowers in full bloom against the blue of the ocean. Exposure to wind, salt spray, and impoverished sandy soil, however, can thwart attempts at a colorful seaside garden. If it is difficult to establish a windbreak (see Windy Sites, page 45), then a sunken garden may be the answer. In either case, choose flowering plants that tolerate exposed, windy conditions and salt spray—such as sea-pink, sweet alyssum, coreopsis, and gaillardia.

*Coastal gardens require plants that can endure wind and salt spray. Lilies, yarrow, poppies, and cosmos bloom successfully beside a harbor in Maine.*

Sandy soil can be an excellent growing medium when supplemented with fertilizer and organic matter. Dairy farms and horse farms often abound by the seashore, so a truckload of manure may be easily available and inexpensive. Seaweed collected from the shoreline also makes excellent compost; simply heap it into piles and mix in garden or kitchen waste. Spraying both sides of leaves with a liquid fertilizer (foliar feeding) will also supply nutrients missing from coastal soils.

## DESERT SOIL

In regions of low rainfall, the soil is often highly alkaline and therefore inhospitable to many plants. Plants growing in alkaline soil frequently show signs of chlorosis, a yellowing of leaves caused by lack of iron. (Iron exists in these soils but in insoluble form.) Although slightly alkaline soils can be improved by adding compost, peat moss, or commercial soil sulfur, trying to amend extremely alkaline soils is often pointless, since surrounding alkaline salts may quickly leach into the amendment and render the soil alkaline again.

What's more, an impervious calcium crust called caliche often forms just below desert topsoil, making gardening even more of a challenge. If the caliche is thin enough, you can chip holes or "chimneys" through it with a pickax or a crowbar. However, even if you can break through it, the soil beneath may still be too alkaline for plants. The easiest remedy is to make raised beds using imported topsoil. Build a frame of landscape ties and fill it with at least 2 feet of soil to provide drainage and isolation from the alkaline layer.

In desert regions, where water is scarce and often rationed, make a reservoir to hold household waste water for recycling into the garden. Dump water from baths and dishwashing into a special metal tank, a pool disguised as a lily pond, or a rain barrel. If you raise the tank or barrel above the level of the flower garden, you won't have to buy a pump; a hose will carry water to the garden by gravity. Some drip systems (see page 34) will work from a gravity-fed water source. There are many other ingenious ways to trap water. In Bermuda, where rainfall is precious, not only do the rooftops all drain into underground reservoirs, but pathways and even the surfaces of tennis courts have been designed to collect rainwater.

*A desert garden takes advantage of a road cut, which has broken up hard, crusty soil, allowing petunias to grow along its edge. Soil behind a retaining wall may be supplemented with compost to hold moisture and neutralize alkalinity.*

## INSECTS, PESTS, AND DISEASES

A flower garden stocked with plants recommended in this book will be naturally resistant to most pests and diseases. However, not even the sturdiest plants are completely immune to attack.

The most effective precaution you can take against garden pests is to keep your plants healthy. Plants that receive adequate water and nutrients and are grown in decent soil tend to resist disease and damage by insects. Keeping the garden neat is another wise precaution. At the end of the growing season, gather up all spent vines, dried stems, and other garden debris for composting or burning. An annual cleanup prevents many insects and disease organisms from overwintering in dead plant material.

If you discover damage you can't diagnose, ask a gardening neighbor or a local garden supplier about likely culprits in your area. For example, in regions with high rainfall, slugs and snails can be a nuisance, especially among new transplants or young seedlings.

Because young seedlings are especially vulnerable, you can avoid many problems by buying plants ready grown. If you do grow plants from seed, you may need to use snail and slug

baits as well as granular insecticides to protect seedlings and young plants. To keep birds, rodents, deer, and other marauders away from flower plantings, spray on chemical repellents or put up barriers such as chicken wire.

Although it is impossible to guard against every insect pest, an effective general-purpose spray is a good start toward pest control. If you prefer organic pesticides, rotenone and pyrethrum, used in combination, will provide a broad control, as will insecticidal soaps. Longer-lasting and usually stronger chemical formulations for general-purpose spraying include diazinon, Sevin®, malathion, and Orthene®. These chemicals are the active ingredients in a variety of products sold at garden centers.

Plant diseases caused by molds, bacteria, and viruses are often hard to control once they gain a foothold. Apart from practicing good garden hygiene, the most effective way to prevent diseases from becoming a problem is to choose disease-resistant cultivars. For example, the Ruffles series of hybrid zinnias and the 'Goldsturm' variety of black-eyed-susan are both resistant to powdery mildew. If a fungal disease does take hold, try a commercial fungicide with a label that says it is effective for the particular disease and plant.

*Slugs feed voraciously in wet regions and can quickly strip plants such as this zinnia. Control them with slug bait or remove them by hand.*

*Hostas with decorative leaves are the mainstay of this shade garden. Shade-tolerant wax begonias have been planted as an edging.*

## SHADE

There are many kinds of shade—morning shade, noontime shade, afternoon shade, filtered shade beneath high trees, deep shade at the foot of low trees or high walls. Few flowering plants will do well in deep shade (impatiens, coleus, wax begonia, and periwinkle are among the most shade loving), but a surprising number will thrive in light shade or in a bed that is shaded for part of the day. For example, tulips, which grow wild in the sunny, arid Middle East, will actually flower longer if planted in light shade.

In a very shady garden, introducing just a little more light can greatly increase the variety of flowers that will grow there. Is your garden deeply shaded by trees? If so, removing a single tree limb may solve the problem.

Does an eave of the house shade the planting area at a critical time of day—perhaps at noon? If it does, and you cannot move the planting site, consider making the surrounding surfaces more reflective. Laying white landscape chips over the soil or painting a nearby fence white can boost light intensity just enough to make the difference, especially for flowering shrubs such as roses.

## SHORT GROWING SEASON

Any place with 100 or fewer days a year of frost-free weather is considered to have a short growing season. Gardens in high altitudes and high latitudes have the shortest growing seasons, yet even in these areas, frost-sensitive flowers can be grown if the soil is loose and fertile. Many flowering plants (such as triploid hybrid marigolds and hybrid zinnias) flower within 40 to 60 days of sowing seed.

You can extend the growing season considerably by planting flowers in cold frames or in raised beds that you can cover quickly when frosty nights threaten. Cold frames usually have hinged tops made of glass or clear plastic, which can be easily opened and closed. The frames are often sunk into soil that has been excavated 2 to 3 feet deep, with just a few inches of the frame projecting above the surface. This form of sunken greenhouse insulates young plants from frigid air while allowing them room to grow upward.

Plants in raised beds can be protected by draping clear plastic sheets or lightweight floating row covers over the beds each evening. Floating row covers are made of white gauze that provides enough insulation to prevent damage from light frost—thus extending the growing season by a month or more.

High-altitude gardens must also contend with the additional problem of cool nights during the growing season. The solution to this problem is to select only cool-season plants that can survive mild frosts, and to set these out as transplants ready to flower. Some good plants for high-altitude gardens are dahlia, columbine, snapdragon, calendula, sweet alyssum, and marigold.

## SLOPES AND HILLTOPS

Gardening on hillsides can be difficult because of their exposure to the elements. Sunny south-facing slopes, for example, are often exceedingly dry and north-facing slopes may be exceptionally shady, limiting the variety of plants that will grow there. Without plants to properly anchor the soil, erosion from wind and rain can also be a problem. The solution lies in careful landscaping and in selecting plants that have sturdy roots and growth habits to gird them against the elements.

Gentle slopes are relatively easy to turn into rock gardens, using boulders and natural-looking rocks to stabilize the soil. Use stones also to form planting pockets for tenacious plants such as stonecrop and desert-candle. It helps if your plants are drought tolerant, since moisture drains away quickly from hillsides and access for watering may be difficult.

Before planting, observe how the hillside drains; avoid placing flower beds where runoff

will wash away the plants during heavy rains. If the hillside is completely devoid of trees, try planting evergreens at random spacings to act as buffers from swirling wind (see Windy Sites, page 45); leave space between them for flower beds.

Steep slopes present more of a challenge. Consider planting the entire slope with a carefree vine, such as Japanese honeysuckle. First, though, you'll need to terrace the slope to contain erosion and create flat areas for planting. Terraces are usually built with stone or with less-expensive landscape ties. Place drainage spouts at the base of each terrace to keep water from building up and dislodging the walls, and if possible, join the terraces together with steps to make garden maintenance easier.

*Top: Sunken cold frames protect plants in areas with short growing seasons. A technique called plunging provides winter refuge for established perennials in pots.*
*Bottom: Terraces built with landscape ties create flat planting spaces on a slope.*

Cascading plants such as sweet alyssum, English lavender, and moss phlox look magnificent spilling over terrace edges or emerging from cracks in stone walls. For shady terraces and walls, choose periwinkle.

Hilltops are equally difficult places to grow flowers because of their exposure to storms and winds. When planting on a hilltop, consider sinking the garden several feet below the crest of the hill for shelter.

## SOGGY SOIL

Gardens with areas of low-lying terrain, heavy clay soil, or impervious subsoil are often subject to waterlogging, which can severely limit the variety of flowering plants that will thrive there. Water that collects in a low-lying area can be drained with a system of pipes placed at least 2 feet underground. The pipes should channel water to a drain, stream, pond, or other depository. If this sounds like too much work or expense, try raised beds. Put down a layer of crushed stones for drainage, then build the frame of the bed with landscape ties, stone, or brick and fill it with good garden topsoil.

Raised beds are also the easiest way to overcome problems with heavy clay soil or impervious subsoil.

Soil that receives a lot of moisture may be acidic. Highly acid soils lock up nutrients, making them unavailable to plants, although there are certain plants—azaleas, for example—that thrive in acid soil. To modify soil acidity, add lime, a powdery soil amendment sold at garden centers. A sprinkling of lime at the rate of about 1 pound per 30 square feet, applied every three years, is usually sufficient to bring moderately acid soil into the neutral range. If the soil is more than moderately acid, gardening in raised beds with imported topsoil may be the only answer. If you suspect that your soil is highly acid, consult a gardening neighbor or ask at your local garden center for advice on soil testing.

## STONY SOIL

Some stony soils are simply gravelly; others contain large solid rocks. Like sand, gravel doesn't hold moisture or nutrients; yet it is an acceptable growing medium, since most roots

*Multiflora petunias and purple alyssum spill over a rocky retaining wall on a slope.*

*Raised beds filled with imported topsoil are the easy solution in areas with stony or waterlogged soil. They are also useful where there is clay or desert soil.*

*A wind-resistant yew hedge shelters a lawn and bench at the Hidcote Manor garden in England.*

can penetrate it freely and it offers excellent drainage. Regular watering and a feeding program (such as spraying leaves with diluted liquid fertilizer every two weeks) will remedy the deficiencies of gravel and allow gravelly soil to yield spectacular results.

In very stony soil the easiest way to garden is in raised beds. Build them with landscape ties, boulders, or bricks, and fill them with good topsoil purchased through a local nursery or garden center.

## WINDY SITES

Windbreaks offer some protection for plants exposed to the damaging effects of wind. Avoid using fences as windbreaks since a solid barrier can actually magnify air turbulence as wind rushes over it. A more effective alternative is to grow a hedge of wind-resistant plants to cushion the force of the wind for their less-sturdy neighbors. To get the hedge established, build a temporary windbreak, either of burlap sacking nailed to strong posts or of hay bales. Remove the temporary windbreak once the young plants are well established and appear stout enough to survive on their own.

Even more effective than a fence or a living windbreak is a sunken garden. Excavate a circle, rectangle, or other garden shape and place plants below the surface of the surrounding soil.

# Easy Garden Design

*An easy flower garden should have a no-fuss design—yet one that inspires and delights. Here are tips for using flowers creatively in a variety of settings.*

The easiest flower gardens use simple designs but creative flair in the planting, with a skillful interplay of color, texture, and form—vibrant flower colors cooled by background greenery, delicate petals against rough bark, low beds counterbalanced by tall shrubs. A garden need not be large or elaborate to be breathtaking: Striking displays can be achieved in a tiny backyard or with containers on a deck or balcony.

The design you choose will depend on what you have to work with and what you want from the garden. Is your yard flat or sloping, sunny or shaded, dry or damp? Are there expanses of fence or wall that cry out for camouflage? Do children need a grassy area to play in? Would you like extra flowers for cutting? All these needs can be met with garden designs that are easy to implement, easy to maintain, and above all, easy to enjoy.

*A modern design for a backyard garden features plants that have textural interest all year: tall fountaingrass, stonecrop, heather, and desert-candle.*

## FORMAL AND INFORMAL GARDENS

Most garden designs exemplify one of two distinct styles: formal or informal. Formal designs are dominated by symmetrically arranged geometric shapes, such as squares, rectangles, and circles, with boundaries sharply defined by brickwork, tile, or hedging. Within these boundaries the plants are spaced meticulously and are not allowed to spill over the edges. Formal gardens usually borrow from French and Italian Renaissance, Moorish, Persian, or Indian styles of landscaping. All show a liking for orderly reflecting pools, grids of canals, statuary placed at uniform intervals, long vistas ending in an architectural highlight such as a pavilion, and carpetlike beds of plants.

Formal gardens are often designed to complement the architecture of a house, duplicating its structural lines on a horizontal plane. The formal approach is also used in the design of "garden rooms"—spaces within the garden that provide privacy and intimacy.

Informal gardens may have some formal elements, but the plants are allowed to spill over boundaries, lending them a sense of spontaneity. The English cottage garden is the epitome of informal design—an organized chaos where plants are carefully placed to create the illusion of wild abandon. The effect is usually achieved through the harmonious combination of five major plant groups—annuals and flowering bulbs for ground-level sweeps of color; perennials for spontaneous background highlights; and vines, shrubs, and trees to carry color high into the sky. Annual flowers are especially important in cottage gardens because many bloom continuously for up to 10 weeks, providing a foundation of color for short-flowering perennials, trees, and shrubs. Fast-growing climbers such as trumpet-creeper, wisteria, clematis, and climbing rose are often trained on arbors and trellises to camouflage monotonous expanses of fence or wall.

Two other styles of informal garden are the wildflower meadow and the woodland garden. A wildflower meadow is created from seed of native flowers, often sold in mixtures selected for a particular habitat. The easiest method of establishing the flowers is to till islands of soil within the meadow; this way the wildflowers

*Top: This elaborate bed of marigolds at the Conservatory of Flowers in San Francisco is an example of formal symmetry.*
*Center: An informal home garden uses a harmonious balance of tall trees and lower shrubs and hedges.*
*Bottom: This border combines tall cosmos, intermediate-height blue salvia, and dwarf French marigold for an informal, spontaneous look.*

## BEDS AND BORDERS

Most formal garden designs and many informal ones group plants within beds and borders. These groupings may feature plants of a single type, such as flowering bulbs, or combine plants from different groups, such as annuals and shrubs.

A bed is a freestanding area of soil—rectangular, square, round, oval, or kidney-shaped—surrounded by lawn, flagstone, brick, or some other ground cover. It is usually mounded in the middle to improve drainage and to render the plants visible from all sides. The plants are most often arranged concentrically, with the tallest variety in the middle and the shortest around the edge.

A border is a narrow planting along a pathway or at the foot of a hedge, wall, or fence. Borders are often rectangular, but sometimes have an informal, free-flowing edge. Most borders are approached from only one side, by means of a path or lawn area. A very effective planting design is a parallel border, in which two flower borders face each other across a path lined with grass, brick, gravel, flagstone, or wood chips. If the border is against a wall, fence, or hedge, raise the soil at the back of the border so that water drains down from the background feature. Border plants are arranged in tiers, with the tallest at the rear, intermediate-height plants in the middle, and compact plants at the front as an edging.

Some designs call for the plants in a bed or border to be perfectly uniform in height. This

*Left: Lawn surrounds a curved formal bed of coleus edged with lavender-cotton. Right: An informal border of zinnia, crested cockscomb, hollyhock, and chrysanthemum frames a lawn.*

can germinate unhindered by local weeds and grasses. Alternatively, till the entire meadow and scatter the wildflower seeds for a natural look. Wildflower mixtures usually contain a balance of annuals and perennials; the annuals bloom the first season and the perennials take over after that. With luck, some annuals may reseed and flower the next season. After two years, one type of flower sometimes predominates—if indeed any remain at all. Very few wildflowers can compete against aggressive weeds and meadow grasses, especially if the wildflowers are not native to the area.

A woodland garden with many deciduous trees can be a good home for even sun-loving flowers, for many bloom in early spring before trees are in full leaf. Shade-loving flowering shrubs and small trees such as azalea, redbud, and dogwood also look magnificent in woodland gardens. In the cool, light shade surrounding them, many bulbs and perennials flower prolifically—especially daffodils, tulips, bleedingheart, columbine, foxglove, lilies, and hosta.

is the case in so-called carpet bedding, a bedding style in which plants are laid out like a carpet on soil that has been raked flat. A level surface and a uniform plant height are also desirable in traditional parterre gardens, where dwarf hedges and flowering plants delineate circles, diamonds, scrolls, and other formal patterns.

## COLOR CONTINUITY

The most interesting gardens have something coming into bloom at every season. Their beds and borders are a palette of colors that may change throughout the year, yet will remain harmonious. Color schemes can be dramatic, with boldly contrasting hues such as red and yellow, or subdued, with soft pastel shades such as pink and white. Monochromatic plantings—an all-white garden or an all-blue garden, for example—are also effective.

By far the easiest way to design for color is to rely mostly on annuals such as marigolds, wax begonias, and impatiens, which have "perpetual" flowering abilities. They flower all summer, producing a fixed assortment of hues. An alternative technique is to plant a variety of short-flowering perennials, shrubs, and annuals, so that as one fades out of bloom another is coming into bloom. Orchestrating color with this technique requires a carefully timed succession of spring-flowering, summer-flowering, and autumn-flowering plants. Perhaps the most pleasing results are created by combining flowering perennials, shrubs, and trees—which all flower briefly—with long-flowering annuals. That way, beds and borders can display the greatest possible diversity of texture, form, and color—some of it fleeting color and some of it nonstop.

The following table shows the flowering times of all of the plants featured in this book. See the "Plant Selection Guide" (page 67) for instructions on when and how to plant them and how best to use them in the garden.

*A garden of annual marigolds, petunias, and white alyssum provides nonstop color all summer.*

# Blooming Schedule

| Annuals | Mar | Apr | May | Jun | Jul | Aug | Sep | Oct | Nov | Dec |
|---|---|---|---|---|---|---|---|---|---|---|
| *Antirrhinum majus* (snapdragon) | | | ● | ● | ● | ● | ● | ● | | |
| *Begonia × semperflorens* (wax begonia) | | | ● | ● | ● | ● | ● | ● | | |
| *Calendula officinalis* (pot marigold) | | | ● | ● | ● | ● | ● | ● | ● | |
| *Catharanthus roseus* (Madagascar periwinkle) | | | | ● | ● | ● | ● | ● | | |
| *Celosia cristata* (crested cockscomb) | | | | ● | ● | ● | ● | ● | | |
| *Centaurea cyanus* (bachelor's-button) | | | | ● | ● | ● | ● | ● | | |
| *Cleome hasslerana* (spiderflower) | | | | ● | ● | ● | ● | ● | | |
| *Coleus* hybrids (coleus) | | | | ● | ● | ● | ● | ● | | |
| *Consolida ambigua* (larkspur) | | | | ● | ● | | | | | |
| *Cosmos bipinnatus* (cosmos) | | | | | ● | ● | ● | ● | | |
| *Dahlia* hybrids (dahlia) | | | | ● | ● | ● | ● | ● | | |
| *Eschscholzia californica* (California-poppy) | | ● | ● | ● | ● | ● | ● | ● | | |
| *Gazania rigens* (gazania) | | | ● | ● | ● | ● | ● | ● | | |
| *Gomphrena globosa* (globe-amaranth) | | | | | ● | ● | ● | ● | | |
| *Helianthus annuus* (sunflower) | | | | | ● | ● | ● | ● | | |
| *Helichrysum bracteatum* (strawflower) | | | | | ● | ● | ● | ● | | |
| *Impatiens wallerana* (impatiens, patienceplant) | | | ● | ● | ● | ● | ● | ● | | |
| *Ipomoea tricolor* (morning glory) | | | | | ● | ● | ● | ● | | |
| *Lavatera trimestris* (mallow) | | | | ● | ● | ● | ● | ● | | |
| *Lobularia maritima* (sweet alyssum) | | | ● | ● | ● | ● | ● | ● | | |
| *Nicotiana alata* (nicotiana, flowering tobacco) | | | | | ● | ● | ● | | | |
| *Papaver rhoeas* (Shirley poppy) | | | | ● | ● | | | | | |
| *Pelargonium × hortorum* (geranium) | | | ● | ● | ● | ● | ● | ● | | |
| *Petunia* hybrids (petunia) | | | ● | ● | ● | ● | ● | ● | | |
| *Portulaca grandiflora* (portulaca) | | | | ● | ● | ● | ● | ● | | |
| *Rudbeckia hirta* (black-eyed-susan, gloriosa daisy) | | | | | ● | ● | ● | | | |
| *Salvia farinacea* (blue salvia) | | | | | ● | ● | ● | ● | | |
| *Salvia splendens* (scarlet sage) | | | | ● | ● | ● | ● | ● | | |
| *Tagetes patula* (French marigold) | | | ● | ● | ● | ● | ● | ● | | |
| *Thunbergia alata* (black-eyed-susan vine) | | | | ● | ● | ● | ● | ● | | |
| *Tropaeolum majus* (nasturtium) | | | | | | ● | ● | ● | | |
| *Verbena* hybrids (verbena) | | | | | ● | ● | ● | | | |
| *Zinnia elegans* (zinnia) | | | | | ● | ● | ● | ● | | |
| **Perennials** | | | | | | | | | | |
| *Achillea filipendulina* (yarrow) | | | | ● | ● | ● | | | | |
| *Agapanthus africanus* (lily-of-the-Nile) | | | ● | ● | ● | ● | ● | ● | | |

## Blooming Schedule

| Perennials (*continued*) | Mar | Apr | May | Jun | Jul | Aug | Sep | Oct | Nov | Dec |
|---|---|---|---|---|---|---|---|---|---|---|
| *Alcea rosea* (hollyhock) | | | | | ■ | ■ | | | | |
| *Aquilegia* hybrids (columbine) | | | ■ | ■ | | | | | | |
| *Armeria maritima* (thrift) | | | ■ | ■ | | | | | | |
| *Asclepias tuberosa* (butterfly weed) | | | | ■ | ■ | | | | | |
| *Aster novae-angliae* (New England aster) | | | | | | ■ | ■ | ■ | | |
| *Astilbe* × *arendsii* (astilbe, false-spirea) | | | | ■ | ■ | | | | | |
| *Aurinia saxatilis* (basket-of-gold, yellow alyssum) | | ■ | ■ | | | | | | | |
| *Baptisia australis* (false-indigo) | | | ■ | | | | | | | |
| *Chrysanthemum* × *superbum* (Shasta daisy) | | | | ■ | ■ | ■ | | | | |
| *Coreopsis lanceolata* (tickseed) | | | | | ■ | ■ | | | | |
| *Cortaderia selloana* (pampas grass) | | | | | | ■ | ■ | ■ | ■ | ■ |
| *Crocus vernus* (giant crocus) | ■ | ■ | | | | | | | | |
| *Dianthus plumarius* (cottage-pinks) | | | ■ | | | | | | | |
| *Dicentra eximia* (bleedingheart) | | | | ■ | ■ | ■ | | | | |
| *Digitalis purpurea* (foxglove) | | | | | ■ | | | | | |
| *Echinacea purpurea* (purple coneflower) | | | | | | ■ | ■ | | | |
| *Gaillardia* × *grandiflora* (Indian-blanket) | | | | ■ | ■ | | | | | |
| *Gladiolus* × *hortulanus* (gladiolus) | | | | | ■ | ■ | | | | |
| *Helianthus* × *multiflorus* (perennial sunflower) | | | | | ■ | | | | | |
| *Heliopsis helianthoides* (false-sunflower) | | | | | ■ | | | | | |
| *Hemerocallis* hybrids (daylily) | | | | | ■ | ■ | | | | |
| *Hibiscus moscheutos* (hardy hibiscus) | | | | | ■ | ■ | | | | |
| *Hosta* hybrids (hosta, plantain lily) | | | | | ■ | | | | | |
| *Iris ensata* (Japanese iris) | | | | | ■ | | | | | |
| *Iris* × *germanica* (bearded iris) | | | ■ | ■ | | | | | | |
| *Iris sibirica* (Siberian iris) | | | | ■ | | | | | | |
| *Lavandula angustifolia* (English lavender) | | | | ■ | ■ | ■ | | | | |
| *Lilium* hybrids (midcentury hybrid lily) | | | | | ■ | | | | | |
| *Lunaria annua* (moneyplant) | | ■ | ■ | | | | | | | |
| *Lycoris squamigera* (naked-ladies) | | | | | ■ | | | | | |
| *Monarda didyma* (beebalm) | | | | | ■ | | | | | |
| *Myosotis scorpioides* (forget-me-not) | ■ | ■ | | | | | | | | |
| *Narcissus* species & hybrids (daffodil) | ■ | ■ | | | | | | | | |
| *Oenothera pilosella* (sundrops, eveningprimrose) | | | | ■ | ■ | | | | | |
| *Paeonia lactiflora* (herbaceous peony) | | | ■ | ■ | | | | | | |
| *Papaver orientale* (Oriental poppy) | | | | ■ | | | | | | |

# Blooming Schedule

| Perennials (*continued*) | Mar | Apr | May | Jun | Jul | Aug | Sep | Oct | Nov | Dec |
|---|---|---|---|---|---|---|---|---|---|---|
| *Phlox subulata* (moss phlox) | | ▬ | ▬ | | | | | | | |
| *Physostegia virginiana* (obedientplant) | | | | | | ▬ | ▬ | | | |
| *Scabiosa caucasica* (pincushionflower) | | | | ▬ | ▬ | | | | | |
| *Sedum spectabile* (stonecrop) | | | | | | ▬ | ▬ | | | |
| *Stachys byzantina* (lamb's-ears) | | | | | ▬ | ▬ | | | | |
| *Tulipa* species & hybrids (tulip) | | ▬ | ▬ | | | | | | | |
| *Yucca filamentosa* (desert-candle) | | | | ▬ | | | | | | |
| **Shrubs, Trees & Woody Vines** | | | | | | | | | | |
| *Albizia julibrissin* (silk tree) | | | | | ▬ | ▬ | | | | |
| *Amelanchier arborea* (sarvis tree) | | ▬ | | | | | | | | |
| *Campsis radicans* (trumpet-creeper) | | | | | ▬ | ▬ | ▬ | | | |
| *Caryopteris × clandonensis* (blue mist shrub) | | | | | ▬ | ▬ | | | | |
| *Cercis canadensis* (eastern redbud) | | ▬ | ▬ | | | | | | | |
| *Chaenomeles speciosa* (flowering quince) | ▬ | ▬ | | | | | | | | |
| *Clematis paniculata* (sweet autumn clematis) | | | | | | ▬ | | | | |
| *Cornus florida* (flowering dogwood) | | | ▬ | | | | | | | |
| *Cotinus coggygria* (smokebush) | | | | ▬ | | | | | | |
| *Crataegus phaenopyrum* (Washington hawthorn) | | | ▬ | | | | | | | |
| *Forsythia × intermedia* (forsythia) | ▬ | ▬ | | | | | | | | |
| *Hibiscus syriacus* (rose-of-Sharon) | | | | | | ▬ | | | | |
| *Hydrangea paniculata* (peegee hydrangea) | | | | | ▬ | ▬ | | | | |
| *Koelreuteria paniculata* (golden-rain-tree) | | | | ▬ | ▬ | | | | | |
| *Lagerstroemia indica* (crapemyrtle) | | | | | ▬ | ▬ | ▬ | | | |
| *Lonicera sempervirens* (scarlet honeysuckle) | | | ▬ | ▬ | | | | | | |
| *Magnolia grandiflora* (southern magnolia) | | ▬ | ▬ | | | | | | | |
| *Magnolia × soulangiana* (saucer magnolia) | | | ▬ | ▬ | | | | | | |
| *Malus floribunda* (Japanese crabapple) | | | ▬ | ▬ | | | | | | |
| *Polygonum aubertii* (silverfleece vine) | | | | | | | ▬ | | | |
| *Prunus cerasifera* 'Atropurpurea' (pissard plum) | | ▬ | | | | | | | | |
| *Pyrus calleryana* (Bradford pear) | | ▬ | | | | | | | | |
| *Raphiolepis umbellata* (Indian-hawthorn) | | ▬ | ▬ | | | | ▬ | | | |
| *Rhododendron* hybrids ('Stewartstonian' azalea) | | | ▬ | | | | | | | |
| *Rosa rugosa* (rugosa rose) | | | | ▬ | ▬ | | | | | |
| *Syringa vulgaris* (common lilac) | | | ▬ | | | | | | | |
| *Vinca minor* (periwinkle) | ▬ | ▬ | ▬ | | | | | | | |
| *Wisteria floribunda* (Japanese wisteria) | | ▬ | ▬ | | | | | | | |

*This assortment of flowering annuals in containers includes pink wax begonias and pink and red geraniums in tubs, and red geraniums in window boxes. In the moss-lined hanging baskets are red-leaved coleus and pink and red impatiens.*

## CONTAINER PLANTINGS

There are many good reasons to grow flowering plants in containers. The most important is the ability of container plantings to bring color to soilless sites such as sun rooms, decks, terraces, and patios. Even in an ample-sized garden, container plantings can be useful as design accents. A decorative urn overflowing with assorted annuals makes a striking highlight in a flower border; a basket of impatiens suspended from a tree limb extends color high above a shade garden. Moreover, it's always a good idea to have some flowering annuals in pots so that you can place them among ground plantings, not only to fill in bare spots but to add color during periods of sparse flowering.

Many flowering plants will grow successfully in containers—indeed some, such as lily-of-the-Nile, actually prefer to have their roots confined and will flower better than if grown in the garden. Page 58 contains a list of easy-to-grow flowering plants for containers.

Choose containers carefully. Containers made of moisture-holding wood and clay are better than plastic or metal ones, which can overheat the soil on hot days and burn tender feeder roots. Large containers are better than small ones, which dry out too quickly. For hanging planters, wire holders lined with sphagnum moss are superior to solid containers (especially plastic ones) because the nest of sphagnum keeps the soil cool and moist.

Although you can use commercial potting soil, it's a little too light in texture to provide good anchorage for container-grown plants. The most successful outdoor container plantings have a mix of two-thirds commercial potting soil and one-third good garden topsoil.

Efficient drainage is vital for container plantings. Many pots have drainage holes to allow any excess moisture to run out, but you must still line the bottoms of these pots with wire or aggregate (such as broken clay pottery) to prevent the drainage holes from clogging. If a container lacks drainage holes and it is difficult to drill some, create a "drainage field" in the bottom of the container with at least 3 inches of crushed stone and charcoal pieces which will collect the water. Lining the bottom of the container this way is usually necessary with window boxes, which rarely have drainage holes. Don't try it with small pots that lack the space for a drainage field.

On sunny days most container plantings need daily watering, and some hanging baskets may need watering twice a day to prevent wilting. To test for moisture, take a pinch of surface soil and rub it between your fingers. Moist soil will stick; dry soil will fall away. Most container plantings are easy to water with a watering can. The best type of watering can has a long, narrow spout that will poke through foliage to apply water directly to the soil. Also useful is a long-handled watering wand that attaches to the end of a garden hose; it will allow you to water hanging baskets and window boxes that may be difficult to reach with a watering can.

### Choosing Containers

Many shapes and sizes of containers are appropriate for gardening. You'll find a large selection at some garden centers and nurseries, or you can search junk shops, antique stores, and auctions for those rusty iron cauldrons

## An Easy-Care Cutting Garden

Many easy-to-grow flowers that look good in the garden are superb for cutting as well. But here's the dilemma: How do you harvest flowers for indoor use without ruining the outdoor display? In a large garden of annuals this may not be a problem, since frequent cutting actually stimulates many varieties to flower more profusely. But a small garden with many perennials may not be able to keep up with demand.

The answer is a separate cutting garden—a utilitarian row or plot, hidden from view at the side of the house or tucked away with the vegetables in a kitchen garden. The design of a cutting garden is practical above all, with straightforward rows that allow access for weeding, watering, and flower gathering. A list of the best flowers to grow for cutting appears on pages 58 and 59.

Gather flowers only in the early morning or late evening—never at midday when they are stressed by heat. Cut only from robust plants, choosing flowers that are just coming into bloom. Use sharp-bladed scissors or pruning shears to sever the stems at a diagonal, cutting stems longer than needed to allow leeway for arranging. Immediately plunge the cut stems into a pail of clean, tepid water and keep them cool and shaded until you are ready to use them.

Some flowers require special treatment after cutting. The leaves of marigolds, zinnias, calendulas, and Shasta daisies decay rapidly underwater and

should be cut off before immersion. Daffodils should be placed in a pail by themselves, since their sap can harm other flowers. Flowers with milky sap, such as Oriental poppies, should have their stem ends singed over a flame or briefly immersed in boiling water to unplug the bead of sap that blocks water uptake. To enhance water absorption by woody stems such as flowering quince, scrape away the last inch or two of bark and make a crisscross cut in the stem end. Hollow-stemmed flowers such as amaryllis and hollyhock

must have their stems filled with water. To do this, turn each stem upside down and pour in water until it is full; then plug the end with cotton.

Most cut flowers last about a week, but some, such as clematis and southern magnolia, stay at their peak for only a day or two. You can extend the vase life of many cut flowers by immersing them in water up to the flower heads (or to the lower leaves of fuzzy-leaved plants) and storing them in a basement overnight before you arrange them.

and chipped stone farm troughs that lend character to a garden. Here are some tips on containers and plants to fill them with.

**Tubs**  Usually round, these are available in clay, plastic, or wood. Whiskey half-barrels make especially fine tubs, since they have plenty of room to accommodate a group of plants. Use tubs on decks and patios. Tulips, daffodils, and hyacinths look particularly beautiful in tubs.

**Urns**  These are striking on pedestals in formal gardens, or as accents in old-fashioned, informal plantings such as English-style cottage gardens. Terracotta and metal are favorite urn materials. Cascading plants such as petunia and periwinkle work well in urns; their flowering stems spill over the edge. A crest of desert-candle in the center of an urn can be surrounded by companion plants such as impatiens and coleus.

**Strawberry barrels**  These are clay or plastic pots, narrow at the top and wider around the middle, with pockets spaced around the sides to hold plants. Pansies, alyssum, begonias, and coleus are just a few of the attractive flowering plants that will dress up a strawberry barrel.

**Window boxes** These are usually made of wood or metal and attached to the side of a building with brackets. The metal kinds can be used only on the shady side of a house, since strong sunlight overheats them. Dwarf flower varieties have been developed especially for growing in the confined spaces of window boxes; for example, 'Imp' impatiens and 'Floral Carpet' snapdragon. Cascading types of geranium, periwinkle, and petunia are also good choices.

**Hanging baskets** These are often made of plastic or wire. Solid plastic baskets should be used only in shade to avoid overheating. Wire baskets lined with sphagnum moss allow plants to be poked through the sides for a rounded, attractive appearance. Many flower varieties have been bred especially for hanging baskets, including 'Cascade' petunia, 'Balcon' geranium, and 'Futura' impatiens.

## AN EASY GARDEN

Presented here is a design for an easy-care garden filled with plants that are described in this book. The design fits an L-shaped section of yard behind a house, with each square of grid representing a square foot of garden space. You can copy this plan exactly or adapt it to your own taste, garden area, and budget.

The design uses a large variety of flowering annuals for summer-long color, a few perennials for seasonal highlights, and a flowering tree and shrub for height and brief color. Each plant is placed where it will do best; coleus and wax begonias grow in the shade of a flowering dogwood tree, for example, and moisture-loving daylilies stand near the edge of a garden pool.

A brick-paved path and patio divide the garden into two planting areas—one around the house foundation and the other at the back of the yard. Around the house, a tall rose-of-Sharon hedge hides an expanse of foundation; shorter petunias, alyssum, and moss phlox spill over onto the brick paving to soften its edges. A wooden bench sits in a recess of the foundation planting, partially shaded by the dogwood tree.

The back of the garden features a small pool surrounded by brightly colored annuals and perennial daylilies. To the right of the pool, a stepping-stone path flanked by cosmos

and globe-amaranth leads to an arbor with benches. A hedge of tall arborvitae (or other evergreen shrub) provides a cooling backdrop that also screens out distractions.

Most plants for this garden may be bought ready grown, but nasturtium, portulaca, and sweet alyssum are easier to plant directly into the garden from seed. Simple arbors are available at garden centers for as little as $60 (higher cost with seats). The pool, which is only a few inches deep, can be lined with PVC pool fabric. A one- to two-inch layer of wood chips can be applied between plants to keep weeds down. The paths and sitting area may be lined with brick, flagstone, or gravel.

## Garden Plant List

For easy reference, here's an alphabetical list of this garden's flowering plants, showing their common and botanical names. You'll find planting instructions in the "Plant Selection Guide" (page 67); blooming times appear in the charts beginning on page 51.

| COMMON NAME | BOTANICAL NAME |
| --- | --- |
| **Annuals** | |
| Black-eyed-susan | *Rudbeckia hirta* |
| Blue salvia | *Salvia farinacea* |
| Coleus | *Coleus* hybrids |
| Cosmos | *Cosmos bipinnatus* |
| Crested cockscomb | *Celosia cristata* |
| Dahlia | *Dahlia* hybrids |
| French marigold | *Tagetes patula* |
| Geranium | *Pelargonium* × *hortorum* |
| Globe-amaranth | *Gomphrena globosa* |
| Impatiens | *Impatiens wallerana* |
| Nasturtium | *Tropaeolum majus* |
| Petunia | *Petunia* hybrids |
| Portulaca | *Portulaca grandiflora* |
| Scarlet sage | *Salvia splendens* |
| Spiderflower | *Cleome hasslerana* |
| Sweet alyssum | *Lobularia maritima* |
| Wax begonia | *Begonia* × *semperflorens-cultorum* |
| Zinnia | *Zinnia elegans* |
| **Perennials** | |
| Beebalm | *Monarda didyma* |
| Daylily | *Hemerocallis* hybrids |
| Moss phlox | *Phlox subulata* |
| Siberian iris | *Iris sibirica* |
| **Woody Plants** | |
| Flowering dogwood | *Cornus florida* |
| Rose-of-Sharon | *Hibiscus syriacus* |

**Design for an Easy-Care Garden**

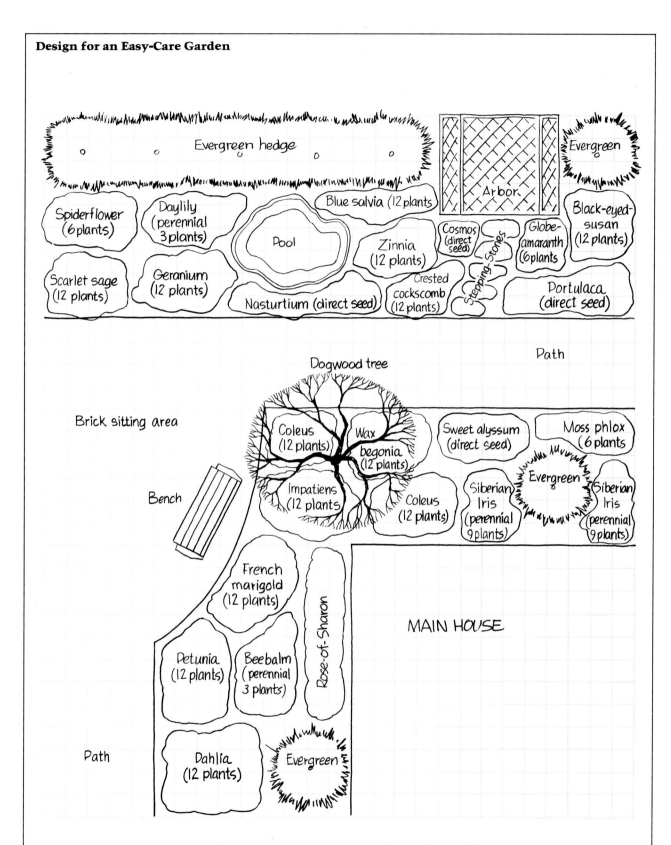

This design for an easy-care backyard flower garden creates a private space that can be enjoyed from many vantage points. Its bench, arbor, and pool form centers of interest accessible by brick and flagstone paths.

## FLOWERS FOR EVERY PURPOSE

One key to successful garden design is choosing the right plants for the right situation. If you can plant flowers whose habits and requirements are closely suited to the conditions in your garden, they will reward you with good health and bountiful flowering.

The following lists identify plants that flower under difficult or specialized conditions. To help you with garden planning, flowers are also listed by color. Note that some plants appear on more than one list. You'll find a complete description of each plant in the next chapter.

*Salvia splendens* and *Impatiens wallerana*

### Climbing and Trailing Flowers

*Campsis radicans*, trumpet-creeper (woody vine)

*Clematis paniculata*, sweet autumn clematis (woody vine)

*Ipomoea tricolor*, morning glory (annual vine)

*Lonicera sempervirens*, scarlet honeysuckle (woody vine)

*Polygenum aubertii*, silverfleece vine (woody vine)

*Thunbergia alata*, black-eyed-susan vine (annual vine)

*Tropaeolum majus*, nasturtium (annual vine)

*Wisteria floribunda*, Japanese wisteria (woody vine)

### Flowers for Containers

*Agapanthus africanus*, lily-of-the-Nile (perennial bulb)

*Antirrhinum majus*, snapdragon (annual)

*Begonia* × *semperflorens-cultorum*, wax begonia (annual)

*Catharanthus roseus*, Madagascar periwinkle (annual)

*Celosia cristata*, crested cockscomb (annual)

*Coleus* hybrids*, coleus (annual)

*Crocus vernus*, giant crocus (perennial bulb)

*Good for hanging baskets

*Dahlia* hybrids, dahlia (annual)

*Digitalis purpurea*, foxglove (perennial)

*Impatiens wallerana**, impatiens, patienceplant (annual)

*Lagerstroemia indica*, crapemyrtle (tree)

*Lilium* hybrids, midcentury hybrid lily (perennial)

*Lobularia maritima**, sweet alyssum (annual)

*Lycoris squamigera*, naked-ladies (perennial)

*Narcissus* species and hybrids, daffodil (perennial bulb)

*Pelargonium* × *hortorum**, geranium (annual)

*Petunia* hybrids*, petunia (annual)

*Salvia farinacea*, blue salvia (annual)

*Salvia splendens*, scarlet sage (annual)

*Sedum spectabile*, stonecrop (perennial)

*Tagetes patula*, French marigold (annual)

*Thunbergia alata**, black-eyed-susan vine (annual vine)

*Tropaeolum majus**, nasturtium (annual vine)

*Tulipa* species and hybrids, tulip (perennial bulb)

*Verbena* hybrids*, verbena (annual)

*Vinca minor**, periwinkle (shrub)

*Zinnia elegans*, zinnia (annual)

### Flowers for Cutting

*Achillea filipendulina*, yarrow (perennial)

*Agapanthus africanus*, lily-of-the-Nile (perennial bulb)

*Antirrhinum majus*, snapdragon (annual)

*Aquilegia* hybrids, columbine (perennial)

*Asclepias tuberosa*, butterfly weed (perennial)

*Aster novae-angliae*, New England aster (perennial)

*Astilbe* × *arendsii*, astilbe, false-spirea (perennial)

*Calendula officinalis*, pot marigold (annual)

*Caryopteris* × *clandonensis*, blue mist shrub (shrub)

*Celosia cristata*, crested cockscomb (annual)

*Centaurea cyanus*, bachelor's-button (annual)

*Chaenomeles speciosa*, flowering quince (shrub)

*Chrysanthemum* × *superbum*, Shasta daisy (perennial)

*Cleome hasslerana*, spiderflower (annual)

*Consolida ambigua*, larkspur (annual)

*Coreopsis lanceolata*, tickseed (perennial)

*Cornus florida*, flowering dogwood (tree)

*Cortaderia selloana,* pampas grass (perennial)

*Cosmos bipinnatus,* cosmos (annual)

*Cotinus coggygria,* smokebush (shrub)

*Dahlia* hybrids, dahlia (annual)

*Digitalis purpurea,* foxglove (perennial)

*Echinacea purpurea,* purple coneflower (perennial)

*Forsythia × intermedia,* forsythia (shrub)

*Gaillardia × grandiflora,* gaillardia, Indian-blanket (perennial)

*Gladiolus × hortulanus,* gladiolus (perennial bulb)

*Gomphrena globosa,* globe-amaranth (annual)

*Helianthus annuus,* sunflower (annual)

*Helianthus × multiflorus,* perennial sunflower (perennial)

*Heliopsis helianthoides* var. *scabra,* false-sunflower (perennial)

*Hydrangea paniculata* 'Grandiflora', peegee hydrangea (shrub)

*Iris × germanica,* bearded iris (perennial)

*Iris sibirica,* Siberian iris (perennial)

*Lagerstroemia indica,* crapemyrtle (tree)

*Lavandula angustifolia,* English lavender (perennial)

*Lilium* hybrids, midcentury hybrid lily (perennial)

*Magnolia × soulangiana,* saucer magnolia (tree)

*Monarda didyma,* beebalm (perennial)

*Narcissus* species and hybrids, daffodil (perennial bulb)

*Paeonia lactiflora,* herbaceous peony (perennial)

*Physostegia virginiana,* obedientplant (perennial)

*Rudbeckia hirta,* black-eyed-susan, gloriosa daisy (annual)

*Salvia farinacea,* blue salvia (annual)

*Scabiosa caucasica,* pincushionflower (perennial)

*Gladiolus*

*Sedum spectabile,* stonecrop (perennial)

*Syringa vulgaris,* common lilac (shrub)

*Tagetes patula,* French marigold (annual)

*Tulipa* species and hybrids, tulip (perennial bulb)

*Zinnia elegans,* zinnia (annual)

## Flowers for Dried Arrangements

*Achillea filipendulina,* yarrow (perennial)

*Armeria maritima,* thrift (perennial)

*Celosia cristata,* crested cockscomb (annual)

*Cortaderia selloana,* pampas grass (perennial)

*Cotinus coggygria,* smokebush (shrub)

*Gomphrena globosa,* globe-amaranth (annual)

*Helichrysum bracteatum,* strawflower (annual)

*Hydrangea paniculata* 'Grandiflora', peegee hydrangea (shrub)

*Koelreuteria paniculata,* golden-rain-tree (tree)

*Lavandula angustifolia,* English lavender (perennial)

*Lunaria annua,* moneyplant (perennial)

*Sedum spectabile,* stonecrop (perennial)

*Stachys byzantina,* lamb's-ears (perennial)

*Rosa rugosa*

## Drought-Resistant Flowers

*Agapanthus africanus*, lily-of-the-Nile (perennial bulb)

*Armeria maritima*, thrift (perennial)

*Aurinia saxatilis*, basket-of-gold, yellow alyssum (perennial)

*Baptisia australis*, false-indigo (perennial)

*Catharanthus roseus*, Madagascar periwinkle (annual)

*Cleome hasslerana*, spiderflower (annual)

*Cortaderia selloana*, pampas grass (perennial)

*Gaillardia* × *grandiflora*, gaillardia, Indian-blanket (perennial)

*Gazania rigens*, gazania (annual)

*Gomphrena globosa*, globe-amaranth (annual)

*Hemerocallis* hybrids, daylily (perennial)

*Iris* × *germanica*, bearded iris (perennial)

*Phlox subulata*, moss phlox (perennial)

*Portulaca grandiflora*, portulaca (annual)

*Raphiolepis umbellata*, Indian-hawthorn (shrub)

*Rosa rugosa*, rugosa rose (shrub)

*Sedum spectabile*, stonecrop (perennial)

*Tagetes patula*, French marigold (annual)

*Yucca filamentosa*, desert-candle (perennial)

## Flowers for Edging

*Armeria maritima*, thrift (perennial)

*Aurinia saxatilis*, basket-of-gold, yellow alyssum (perennial)

*Begonia* × *semperflorens-cultorum*, wax begonia (annual)

*Catharanthus roseus*, Madagascar periwinkle (annual)

*Coleus* hybrids, coleus (annual)

*Crocus vernus*, giant crocus (perennial bulb)

*Dianthus plumarius*, cottage-pinks (perennial)

*Dicentra eximia*, bleedingheart (perennial)

*Eschscholzia californica*, California-poppy (annual)

*Gazania rigens*, gazania (annual)

*Impatiens wallerana*, impatiens, patienceplant (annual)

*Lavandula angustifolia*, English lavender (perennial)

*Lobularia maritima*, sweet alyssum (annual)

*Myosotis scorpioides*, forget-me-not (perennial)

*Petunia* hybrids, petunia (annual)

*Phlox subulata*, moss phlox (perennial)

*Portulaca grandiflora*, portulaca (annual)

*Stachys byzantina*, lamb's-ears (perennial)

*Tagetes patula*, French marigold (annual)

*Verbena* hybrids, verbena (annual)

*Vinca minor*, periwinkle (shrub)

## Ground Covers

*Agapanthus africanus*, lily-of-the-Nile (perennial bulb)

*Armeria maritima*, thrift (perennial)

*Catharanthus roseus*, Madagascar periwinkle (annual)

*Dicentra eximia*, bleedingheart (perennial)

*Gazania rigens*, gazania (annual)

*Hemerocallis* hybrids, daylily (perennial)

*Hosta sieboldiana*, hosta, plantain lily (perennial)

*Impatiens wallerana,* impatiens, patienceplant (annual)

*Lavandula angustifolia,* English lavender (perennial)

*Lobularia maritima,* sweet alyssum (annual)

*Myosotis scorpioides,* forget-me-not (perennial)

*Oenothera pilosella,* sundrops, eveningprimrose (perennial)

*Phlox subulata,* moss phlox (perennial)

*Raphiolepis umbellata,* Indian-hawthorn (shrub)

*Stachys byzantina,* lamb's-ears (perennial)

*Tropaeolum majus,* nasturtium (annual vine)

*Verbena* hybrids, verbena (annual)

*Vinca minor,* periwinkle (shrub)

## Flowers for Hedging

*Caryopteris* × *clandonensis,* blue mist shrub (shrub)

*Cortaderia selloana,* pampas grass (perennial)

*Forsythia* × *intermedia,* forsythia (shrub)

*Hibiscus syriacus,* rose-of-Sharon (shrub)

*Hydrangea paniculata* 'Grandiflora', peegee hydrangea (shrub)

*Lavatera trimestris,* mallow (annual)

*Paeonia lactiflora,* herbaceous peony (perennial)

*Raphiolepis umbellata,* Indian-hawthorn (shrub)

*Rhododendron* hybrids, 'Stewartstonian' azalea (shrub)

*Rosa rugosa,* rugosa rose (shrub)

*Syringa vulgaris,* common lilac (shrub)

## Moisture Lovers

*Astilbe* × *arendsii,* astilbe, false-spirea (perennial)

*Begonia* × *semperflorens-cultorum,* wax begonia (annual)

*Coleus* hybrids, coleus (annual)

*Digitalis purpurea,* foxglove (perennial)

*Hibiscus moscheutos* 'Southern Belle', hardy hibiscus (perennial)

*Hosta sieboldiana,* hosta, plantain lily (perennial)

*Impatiens wallerana,* impatiens, patienceplant (annual)

*Iris ensata,* Japanese iris (perennial)

*Iris sibirica,* Siberian iris (perennial)

*Lunaria annua,* moneyplant (perennial)

*Myosotis scorpioides,* forget-me-not (perennial)

*Vinca minor,* periwinkle (shrub)

## Flowers That Naturalize

*Agapanthus africanus,* lily-of-the-Nile (perennial bulb)

*Armeria maritima,* thrift (perennial)

*Asclepias tuberosa,* butterfly weed (perennial)

*Baptisia australis,* false-indigo (perennial)

*Centaurea cyanus,* bachelor's-button (annual)

*Clematis paniculata,* sweet autumn clematis (woody vine)

*Consolida ambigua,* larkspur (annual)

*Coreopsis lanceolata,* tickseed (perennial)

*Cortaderia selloana,* pampas grass (perennial)

*Cosmos bipinnatus,* cosmos (annual)

*Crocus vernus,* giant crocus (perennial bulb)

*Dicentra eximia,* bleedingheart (perennial)

*Digitalis purpurea,* foxglove (perennial)

*Narcissus*

*Echinacea purpurea*, purple
  coneflower (perennial)
*Eschscholzia californica*, California-
  poppy (annual)
*Hemerocallis* hybrids, daylily
  (perennial)
*Iris × germanica*, bearded iris
  (perennial)
*Iris ensata*, Japanese iris
  (perennial)
*Iris sibirica*, Siberian iris
  (perennial)
*Lilium* hybrids, midcentury hybrid
  lily (perennial)
*Lobularia maritima*, sweet alyssum
  (annual)
*Lunaria annua*, moneyplant
  (perennial)
*Monarda didyma*, beebalm
  (perennial)
*Narcissus* species and hybrids,
  daffodil (perennial bulb)
*Oenothera pilosella*, sundrops,
  eveningprimrose (perennial)
*Papaver rhoeas*, Shirley poppy
  (annual)
*Phlox subulata*, moss phlox
  (perennial)
*Rudbeckia hirta*, black-eyed-susan,
  gloriosa daisy (annual)
*Tulipa* species and hybrids, tulip
  (perennial bulb)

## Shade Lovers

*Agapanthus africanus*, lily-of-the-
  Nile (perennial bulb)
*Aquilegia* hybrids, columbine
  (perennial)
*Astilbe × arendsii*, astilbe, false-
  spirea (perennial)
*Begonia × semperflorens-cultorum*,
  wax begonia (annual)
*Cercis canadensis*, eastern redbud
  (tree)
*Coleus* hybrids, coleus (annual)
*Cornus florida*, flowering dogwood
  (tree)
*Dicentra eximia*, bleedingheart
  (perennial)
*Digitalis purpurea*, foxglove
  (perennial)
*Hosta sieboldiana*, hosta, plantain
  lily (perennial)

*Myosotis scorpioides* (with *Cheiranthus cherri*)

*Impatiens wallerana*, impatiens,
  patienceplant (annual)
*Lilium* hybrids, midcentury hybrid
  lily (perennial)
*Lunaria annua*, moneyplant
  (perennial)
*Myosotis scorpioides*, forget-me-not
  (perennial)
*Narcissus* species and hybrids,
  daffodil (perennial bulb)
*Rhododendron* hybrids,
  'Stewartstonian' azalea (shrub)
*Tulipa* species and hybrids, tulip
  (perennial bulb)
*Vinca minor*, periwinkle (shrub)

## Blue and Lavender Flowers

*Agapanthus africanus*, lily-of-the-
  Nile (perennial bulb)
*Aquilegia* hybrids, columbine
  (perennial)
*Aster novae-angliae*, New England
  aster (perennial)
*Baptisia australis*, false-indigo
  (perennial)

*Caryopteris × clandonensis*, blue
  mist shrub (shrub)
*Centaurea cyanus*, bachelor's-button
  (annual)
*Consolida ambigua*, larkspur
  (annual)
*Crocus vernus*, giant crocus
  (perennial bulb)
*Gladiolus × hortulanus*, gladiolus
  (perennial bulb)
*Hemerocallis* hybrids, daylily
  (perennial)
*Hibiscus syriacus*, rose-of-Sharon
  (shrub)
*Hosta sieboldiana*, hosta, plantain
  lily (perennial)
*Ipomoea tricolor*, morning glory
  (annual vine)
*Iris × germanica*, bearded iris
  (perennial)
*Iris ensata*, Japanese iris (perennial)
*Iris sibirica*, Siberian iris
  (perennial)
*Lavandula angustifolia*, English
  lavender (perennial)

*Myosotis scorpioides*, forget-me-not (perennial)
*Petunia* hybrids, petunia (annual)
*Phlox subulata*, moss phlox (perennial)
*Salvia farinacea*, blue salvia (annual)
*Scabiosa caucasica*, pincushionflower (perennial)
*Syringa vulgaris*, common lilac (shrub)
*Tulipa* species and hybrids, tulip (perennial bulb)
*Verbena* hybrids, verbena (annual)
*Vinca minor,* periwinkle (shrub)
*Wisteria floribunda*, Japanese wisteria (woody vine)

## Pink and Red Flowers

*Albizia julibrissin,* silk tree (tree)
*Alcea rosea*, hollyhock (perennial)
*Antirrhinum majus*, snapdragon (annual)
*Aquilegia* hybrids, columbine (perennial)

*Armeria maritima*, thrift (perennial)
*Aster novae-angliae*, New England aster (perennial)
*Astilbe* × *arendsii*, astilbe, false-spirea (perennial)
*Begonia* × *semperflorens-cultorum*, wax begonia (annual)
*Catharanthus roseus*, Madagascar periwinkle (annual)
*Celosia cristata*, crested cockscomb (annual)
*Centaurea cyanus*, bachelor's-button (annual)
*Cercis canadensis*, eastern redbud (tree)
*Chaenomeles speciosa*, flowering quince (shrub)
*Cleome hasslerana*, spiderflower (annual)
*Coleus* hybrids, coleus (annual)
*Consolida ambigua*, larkspur (annual)
*Cornus florida*, flowering dogwood (tree)
*Cosmos bipinnatus*, cosmos (annual)
*Cotinus coggygria*, smokebush (shrub)
*Crataegus phaenopyrum*, Washington hawthorn (tree)
*Dahlia* hybrids, dahlia (annual)
*Dianthus plumarius*, cottage-pinks (perennial)
*Dicentra eximia*, bleedingheart (perennial)
*Digitalis purpurea*, foxglove (perennial)
*Echinacea purpurea*, purple coneflower (perennial)
*Gaillardia* × *grandiflora*, gaillardia, Indian-blanket (perennial)
*Gazania rigens*, gazania (annual)
*Gladiolus* × *hortulanus*, gladiolus (perennial bulb)
*Helichrysum bracteatum*, strawflower (annual)
*Hemerocallis* hybrids, daylily (perennial)
*Hibiscus moscheutos* 'Southern Belle', hardy hibiscus (perennial)
*Hibiscus syriacus*, rose-of-Sharon (shrub)

*Impatiens wallerana*, impatiens, patienceplant (annual)
*Ipomoea tricolor,* morning glory (annual vine)
*Iris* × *germanica*, bearded iris (perennial)
*Lagerstroemia indica*, crapemyrtle (tree)
*Lavatera trimestris*, mallow (annual)
*Lilium* hybrids, midcentury hybrid lily (perennial)
*Lobularia maritima*, sweet alyssum (annual)
*Lonicera sempervirens*, scarlet honeysuckle (woody vine)
*Lunaria annua*, moneyplant (perennial)
*Lycoris squamigera*, naked-ladies (perennial bulb)
*Magnolia* × *soulangiana*, saucer magnolia (tree)
*Malus floribunda*, Japanese crabapple (tree)
*Monarda didyma*, beebalm (perennial)
*Myosotis scorpioides*, forget-me-not (perennial)
*Nicotiana alata*, nicotiana, flowering tobacco (annual)
*Paeonia lactiflora*, herbacious peony (perennial)
*Papaver orientale*, Oriental poppy (perennial)
*Papaver rhoeas*, Shirley poppy (annual)
*Pelargonium* × *hortorum*, geranium (annual)
*Petunia* hybrids, petunia (annual)
*Phlox subulata*, moss phlox (perennial)
*Physostegia virginiana*, obedientplant (perennial)
*Portulaca grandiflora*, portulaca (annual)
*Raphiolepis umbellata*, Indian-hawthorn (shrub)
*Rhododendron* hybrids, 'Stewartstonian' azalea (shrub)
*Rosa rugosa*, rugosa rose (shrub)
*Salvia splendens*, scarlet sage (annual)

*Sedum spectabile,* stonecrop (perennial)

*Syringa vulgaris,* common lilac (shrub)

*Tagetes patula,* French marigold (annual)

*Tropaeolum majus,* nasturtium (annual vine)

*Tulipa* species and hybrids, tulip (perennial bulb)

*Verbena* hybrids, verbena (annual)

*Zinnia elegans,* zinnia (annual)

## Yellow and Orange Flowers

*Achillea filipendulina,* yarrow (perennial)

*Alcea rosea,* hollyhock (perennial)

*Antirrhinum majus,* snapdragon (annual)

*Aquilegia* hybrids, columbine (perennial)

*Asclepias tuberosa,* butterfly weed (perennial)

*Aurinia saxatilis,* basket-of-gold, yellow alyssum (perennial)

*Calendula officinalis,* pot marigold (annual)

*Campsis radicans,* trumpet-creeper (woody vine)

*Celosia cristata,* crested cockscomb (annual)

*Chaenomeles speciosa,* flowering quince (shrub)

*Coleus* hybrids, coleus (annual)

*Coreopsis lanceolata,* tickseed (perennial)

*Dahlia* hybrids, dahlia (annual)

*Digitalis purpurea,* foxglove (perennial)

*Eschscholzia californica,* California-poppy (annual)

*Forsythia × intermedia,* forsythia (shrub)

*Gaillardia × grandiflora,* gaillardia, Indian-blanket (perennial)

*Gazania rigens,* gazania (annual)

*Gladiolus × hortulanus,* gladiolus (perennial bulb)

*Gomphrena globosa,* globe-amaranth (annual)

*Helianthus annuus,* sunflower (annual)

*Helianthus annuus*

*Helianthus × multiflorus,* perennial sunflower (perennial)

*Helichrysum bracteatum,* strawflower (annual)

*Heliopsis helianthoides* var. *scabra,* false-sunflower (perennial)

*Hemerocallis* hybrids, daylily (perennial)

*Iris × germanica,* bearded iris (perennial)

*Koelreuteria paniculata,* golden-rain-tree (tree)

*Lilium* hybrids, midcentury hybrid lily (perennial)

*Narcissus* species and hybrids, daffodil (perennial bulb)

*Nicotiana alata,* nicotiana, flowering tobacco (annual)

*Oenothera pilosella,* sundrops, eveningprimrose (perennial)

*Petunia* hybrids, petunia (annual)

*Portulaca grandiflora,* portulaca (annual)

*Rudbeckia hirta,* black-eyed-susan, gloriosa daisy (annual)

*Tagetes patula,* French marigold (annual)

*Thunbergia alata,* black-eyed-susan vine (annual vine)

*Tropaeolum majus,* nasturtium (annual vine)

*Tulipa* species and hybrids, tulip (perennial bulb)

*Zinnia elegans,* zinnia (annual)

## White Flowers

*Agapanthus africanus,* lily-of-the-Nile (perennial bulb)

*Alcea rosea,* hollyhock (perennial)

*Amelanchier arborea,* sarvis tree (tree)

*Antirrhinum majus,* snapdragon (annual)

*Aquilegia* hybrids, columbine (perennial)

*Aster novae-angliae,* New England aster (perennial)

*Astilbe × arendsii,* astilbe, false-spirea (perennial)

*Begonia × semperflorens-cultorum,* wax begonia (annual)

*Catharanthus roseus,* Madagascar periwinkle (annual)

*Centaurea cyanus*, bachelor's-button (annual)

*Cercis canadensis*, eastern redbud (tree)

*Chrysanthemum × superbum*, Shasta daisy (perennial)

*Clematis paniculata*, sweet autumn clematis (vine)

*Cleome hasslerana*, spiderflower (annual)

*Consolida ambigua*, larkspur (annual)

*Cornus florida*, flowering dogwood (tree)

*Cortaderia selloana*, pampas grass (perennial)

*Cosmos bipinnatus*, cosmos (annual)

*Crataegus phaenopyrum*, Washington hawthorn (tree)

*Crocus vernus*, giant crocus (perennial bulb)

*Dahlia* hybrids, dahlia (annual)

*Dianthus plumarius*, cottage-pinks (perennial)

*Dicentra eximia*, bleedingheart (perennial)

*Digitalis purpurea*, foxglove (perennial)

*Gazania rigens*, ganzania (annual)

*Gladiolus × hortulanus*, gladiolus (perennial bulb)

*Gomphrena globosa*, globe-amaranth (annual)

*Hibiscus moscheutos* 'Southern Belle', hardy hibiscus (perennial)

*Hibiscus syriacus*, rose-of-Sharon (shrub)

*Hosta sieboldiana*, hosta, plantain lily (perennial)

*Hydrangea paniculata* 'Grandiflora', peegee hydrangea (shrub)

*Impatiens wallerana*, impatiens, patienceplant (annual)

*Ipomoea tricolor*, morning glory (annual vine)

*Iris × germanica*, bearded iris (perennial)

*Iris ensata*, Japanese iris (perennial)

*Iris sibirica*, Siberian iris (perennial)

*Lagerstroemia indica*, crapemyrtle (tree)

*Lavandula angustifolia*, English lavender (perennial)

*Lavatera trimestris*, mallow (annual)

*Lilium* hybrids, midcentury hybrid lily (perennial)

*Lobularia maritima*, sweet alyssum (annual)

*Magnolia grandiflora*, southern magnolia (tree)

*Magnolia × soulangiana*, saucer magnolia (tree)

*Narcissus* species and hybrids, daffodil (perennial bulb)

*Nicotiana alata*, nicotiana, flowering tobacco (annual)

*Paeonia lactiflora*, herbaceous peony (perennial)

*Papaver orientale*, Oriental poppy (perennial)

*Pelargonium × hortorum*, geranium (annual)

*Petunia* hybrids, petunia (annual)

*Phlox subulata*, moss phlox (perennial)

*Physostegia virginiana*, obedientplant (perennial)

*Polygonum aubertii*, silverfleece vine (woody vine)

*Portulaca grandiflora*, portulaca (annual)

*Prunus cerasifera* 'Atropurpurea', purple-leaf plum, pissard plum (tree)

*Pyrus calleryana*, Bradford pear (tree)

*Salvia farinacea*, blue salvia (annual)

*Scabiosa caucasica*, pincushionflower (perennial)

*Stachys byzantina*, lamb's-ears (perennial)

*Syringa vulgaris*, common lilac (shrub)

*Tulipa* species and hybrids, tulip (perennial bulb)

*Verbena* hybrids, verbena (annual)

*Vinca minor*, periwinkle (shrub)

*Wisteria floribunda*, japanese wisteria (woody vine)

*Yucca filamentosa*, desert-candle (perennial)

*Zinnia elegans*, zinnia (annual)

*Magnolia grandiflora*

# Plant Selection Guide

*These star performers don't just look good—they are survivors. This guide presents the information you need to select the flowering annuals, perennials, shrubs, trees, and woody vines that are best for your garden.*

A real star makes a performance look easy; the flowering plants in this guide have been chosen above all for that star quality—they will put on a good show while standing up to heat, cold, pests, diseases, poor soil, or drought. Some are North American natives, born and bred to the conditions that prevail here: notably hot summers with extended dry spells. Others evolved or were bred under similar conditions elsewhere. Not all plants listed here will thrive everywhere, but all are true survivors with a high tolerance for adversity.

The listings include traditional favorites, such as French marigold (*Tagetes patula*) and trumpet-creeper (*Campsis radicans*), and also lesser-known plants, such as false-indigo (*Baptisia australis*) and scarlet honeysuckle (*Lonicera sempervirens*), that are well worth discovering. All are widely available, either locally or through mail-order sources such as those listed on page 109.

Sometimes a listing recommends a species, such as *Antirrhinum majus* (snapdragon), and at other times it suggests a specific variety, such as *Ipomoea tricolor* 'Heavenly Blue' ('Heavenly Blue' morning glory), that is either the most widely grown or that stands out most on its merits. Note that plant descriptions give average mature heights at the time of flowering. In fertile soils and in areas with high rainfall, plants may grow taller.

*Easy-care flowers along this stone path include perennial hollyhocks and tickseed, also annual petunias, gloriosa daisies, China asters, and cosmos.*

## *Easiest of the Easy*

All of the flowering plants described on these pages are easy to grow, but some are outstandingly so. The 25 plants listed below are not only virtually trouble free, but also rank as first-rate ornamentals. An all-star garden of these top performers will produce superb nonstop flowering from early spring until fall frost, with a minimum of care and fuss. If you are unfamiliar with any of these, make an effort to try them. You'll be delighted with the result.

### Annuals
*Begonia* × *semperflorens-cultorum*
    (wax begonia)
*Cleome hasslerana* (spiderflower)
*Cosmos bipinnatus* (cosmos)
*Impatiens wallerana* (impatiens,
    patienceplant)
*Lavatera trimestris* (mallow)
*Rudbeckia hirta* (black-eyed-susan,
    gloriosa daisy)
*Tagetes patula* (French marigold)
*Zinnia elegans* (zinnia)

### Perennials
*Chrysanthemum* × *superbum* (Shasta
    daisy)
*Coreopsis lanceolata* (tickseed)
*Echinacea purpurea* (purple
    coneflower)
*Hemerocallis* hybrids (daylily)
*Hibiscus moscheutos* 'Southern Belle'
    (hardy hibiscus)
*Iris sibirica* (Siberian iris)
*Lavandula angustifolia* (English
    lavender)
*Monarda didyma* (beebalm)
*Narcissus* species and hybrids (daffodil)
*Sedum spectabile* (stonecrop)

### Shrubs, Small Trees, and Woody Vines
*Amelanchier arborea* (sarvis tree)
*Campsis radicans* (trumpet-creeper)
*Clematis paniculata* (sweet autumn
    clematis)
*Lonicera sempervirens* (scarlet
    honeysuckle)
*Polygonum aubertii* (silverfleece vine)
*Pyrus calleryana* (Bradford pear)
*Vinca minor* (periwinkle)

*Top to bottom: cosmos, marigolds and
zinnias, daffodils.*

*Antirrhinum majus*

*Begonia × semperflorens-cultorum*

## ANNUALS

Use annuals to fill beds and borders with brilliant, long-lasting color. Most live only one season, but bloom for many weeks. Many will grow readily from seed.

### *Antirrhinum majus*
Snapdragon

Although snapdragons are not heat resistant and usually stop blooming when days turn hot and humid, they make up for this with spectacular flowering during the cool spring months. Moreover, if you cut the flowering stems to just above the soil in midsummer, they'll bloom again in fall.

The classic snapdragon is a spire up to 4 feet tall, with tubular flowers studded along the top third of the stem. Colors include white, yellow, orange, red, pink, and a range of bicolors. Snapdragons are hardy annuals and can be transplanted outdoors several weeks before the last frost date. Plant them in sunny, well-drained soil, and water whenever the soil feels dry.

Snapdragon varieties are classified by height, in tall, medium, and dwarf categories.

The tall Butterfly hybrids (3 feet) have open throats and the most intense coloring, since more of the flower is exposed. Rocket hybrids (3 feet) have the more traditional closed, "dragon" mouth. These two types are the best for cutting, but they may need staking. Medium varieties, such as the 'Coronettes' (2 feet), form bushy clumps that usually need no staking yet produce stems long enough to cut. Dwarf varieties, such as 'Floral Carpet' (8 inches), grow in low mounds that are hard to recognize as snapdragons. They are used mostly in beds and borders.

Snapdragons are easy to grow from seed, but need starting indoors to produce eight-week-old transplants capable of withstanding mild frosts. Nurseries generally offer a good, inexpensive selection of seedlings in six-packs.

### *Begonia × semperflorens-cultorum*
Wax begonia

Hybridizing has improved wax begonias so much that it is no longer worth planting the nonhybrid standard varieties. Whereas standard varieties burn up in the sun, hybrids thrive in both sun and shade; they also bloom more profusely over a longer period. Their only demand is cool, moist soil with a high humus content in the form of peat, compost, or leaf mold.

Native to tropical South America, wax begonias grow throughout the United States and young plants are easy to find at any garden center. (Don't bother with seeds—they are tiny, like specks of dust, and take at least 10 weeks of pampering to sprout and grow to transplantable size.) Owing to their tropical origin, wax begonias must not be transplanted into the garden until frost danger has passed; and the soil around them should be kept moist, as this helps to maintain the cool soil temperatures they need. When watering, take care to avoid splashing the leaves, which induces leaf spots.

Mature plants grow in mounds and cover themselves with a profusion of small, 1-inch flowers that have a large upper and lower petal and two small side petals surrounding a powdery yellow cluster of stamens. Colors include white, red, pink, rose, and bicolors. There is also a choice of leaf colors—green or bronze. Heights range from 12 to 24 inches, depending on variety.

A popular dwarf hybrid is the bronze-leaved Cocktail series, whose members have names like 'Vodka' (scarlet), 'Whiskey' (white), and 'Gin'(rose pink). Dwarf hybrids are excellent for massing in low beds and for edging flower borders. Try them in tubs and window boxes, too. The Glamour series grows tall and bushy, making a solid backdrop for borders.

Calendula officinalis

Catharanthus roseus

Celosia cristata 'Floradale'

### Calendula officinalis
Pot marigold

Calendula seed germinates so reliably that it can be sown on top of snow to sprout in bare soil as the snow melts. It grows quickly and flowers within six weeks of germination. Calendula lacks heat resistance, but otherwise it is extremely hardy, flowering strongly and continuously during cool weather in spring and autumn. Its light green, spicy-smelling, indented foliage with 4-inch densely petalled flowers make a lively bed or border of sunny colors.

Calendula flowers are orange, yellow, apricot, or white, sometimes with a conspicuous dark eye at the center. If flowers dwindle during midsummer, trim back the plants to within 6 inches of the soil line to stimulate repeat

flowering in autumn. For extra-early flowers in spring, sow seed into bare soil in late summer; it will produce stocky plants that can survive a freezing winter.

Dwarf calendulas, such as 'Fiesta Gitana' ('Gypsy Festival'), grow just 12 inches high and will flower indoors in pots during winter. Grow tall kinds, such as 'Pacific Beauty' (2½ feet), in cutting gardens for arrangements.

### Catharanthus roseus
Madagascar periwinkle

Few flowering annuals have the cast-iron constitution of Madagascar periwinkle, a heat-resistant, ground-hugging plant from India. The tough, spear-shaped leaves are so attractively glossy that they look evergreen. The 1½-inch star-shaped flowers appear all summer, undeterred by air pollution or drought.

Colors include white, pink, and purple, some with a contrasting eye that is usually red (on white flowers) or white (on pink or purple flowers). The creeping types, used mostly for ground cover, grow

less than 6 inches high and are the most desirable for a dense floral display. More upright types, up to 1½ feet high, have been developed for container plantings.

Although seeds germinate reliably, the seedlings are susceptible to overwatering and temperature fluctuation. It's better to buy young plants in six-packs for transplanting.

### Celosia cristata
Crested cockscomb

The flowers of crested cockscomb are among the most bizarre of all annuals, resembling clumps of brain coral. Sow seeds directly into the garden after spring frost; they will sprout within seven days and grow rapidly if the flower garden is in full sun. Keep the plantings watered so as not to

weaken the plants, and they will bloom within eight weeks. Crested cockscombs thrive in summer heat; they will bloom throughout the summer if you remove the faded flowers. Deadheading stimulates the plant to produce side branches, which produce more flowers.

Crested cockscombs are yellow, cream, scarlet, pink, or crimson. The spear-shaped leaves are bright green. Plants grow 6 to 24 inches tall, depending on variety. Use dwarf kinds for containers and low bedding, taller kinds in borders or for cutting. The 'Toreador' variety has enormous flower heads, up to 12 inches across, and is valued for dried decorations.

A related species, C. plumosa, has feathery plumes and is equally heat tolerant. 'Forest Fire Improved' is a particularly attractive variety, with bronze foliage and shimmering red plumes.

*Centaurea cyanus*

*Cleome hasslerana*

*Coleus* hybrid 'Flame Nettle'

### *Centaurea cyanus*
Bachelor's-button

So eager is bachelor's-button to bloom that seed sown in the garden will sprout within days and flower within eight weeks. The slender, upright stems are topped by round, flat blue flowers typically 1½ inches across. Other flower colors include blue, pink, white, and maroon. The leaves are gray-green, slender, and lancelike.

Bachelor's-button flowers best when nights are cool. It thrives even in poor soils, and it prefers full sun. Native to the meadows and cornfields of Europe, bachelor's-button is popular in wildflower mixtures, making a particularly attractive meadow when

planted with Shirley poppies (*Papaver rhoeas*). Once you have it in your garden, it is likely to stay, since in fall the seed heads spill new seeds onto the ground to germinate in early spring. Plants usually grow 3 feet high, although some 18-inch dwarf types are available.

Bachelor's-buttons make good lapel flowers, but their stems wilt quickly, limiting their appeal as cut flowers. 'Blue Diadem', developed by an English seed house, has the largest flowers—up to 2 inches across.

### *Cleome hasslerana*
Spiderflower

Few flowering annuals have both the height and the heat resistance of the spiderflower. While foxglove, delphinium, and other tall flowers dwindle as summer days turn hot, this native of arid South America goes on enthusiastically, adding height and flowers to its central stem and side branches as the lower blooms fade. Moreover, it tolerates poor soil as well as glaring sunlight.

The trumpet blossoms are arranged in a crown, with long curving stamens that give the flower head a distinctly spidery appearance. Choose pink, purple, or white—the latter is a favorite plant in all-white theme gardens. The 4- to 5-foot plants look best at the back of a border, in either mixtures or clusters of color. Arrangers like the flowers for large, dramatic displays.

Spiderflowers are easy to grow from seed started indoors, but six-packs of seedlings are inexpensive. Take care, however, to buy compact plants since those that are stretched or already flowering may suffer transplant shock and never achieve strong, uniform growth.

### *Coleus* hybrids
Coleus

The small blue flowers of coleus are insignificant beside the vibrant leaves that are this plant's chief ornamental asset. Although often regarded as an indoor plant, coleus is easy to grow outdoors. It is valued highly for its shade tolerance, although it will grow in sun if the soil is high in moisture-retentive humus and if it is watered when dry.

Coleus seed is tiny, needs high temperature (80° F) to germinate, and takes 10 weeks to reach transplantable size. It is thus far easier to buy flats or six-packs for transplanting into the garden after frost.

At one time the only coleus sold in seed form was Rainbow mixed colors, which tended to bolt to seed and become ungainly by midsummer unless its flower spikes were constantly removed by hand. Now, plant breeders offer the Wizard series, similar in appearance to the famous Rainbow

*Consolida ambigua*

*Cosmos bipinnatus* 'Radiance'

*Dahlia*

but with a new quality called "self-branching." Wizard coleus sprouts side branches low to the ground, producing attractive bushiness, and holds off flowering until extremely late in the season. Both Rainbow and Wizard have colorful, wavy leaves in lime green, red, bronze, yellow, orange, and cream. There may be up to three colors per leaf, with no two leaves alike.

Coleus is sensational massed in low beds, especially in combination with wax begonias and impatiens, which are also shade lovers. Certain varieties are suitable for tubs and hanging baskets, including 'Fiji', which has lacy leaf edges and cascading branches.

### *Consolida ambigua*
Larkspur

Like many other easy-to-grow annuals, larkspur is a cool-weather plant that blooms early in the season and then dwindles as summer temperatures rise. Grown easily from seed sown directly into the garden—even before the last spring frosts—it sprouts like a weed and blooms magnificently for several weeks in early summer. For even earlier blooms, sow seed into the garden in late summer; these hardy plants will produce enough growth to survive the winter and bloom spectacularly the next spring.

Larkspurs are tall, stately plants, 3 to 4 feet high, with lovely delphiniumlike flowers studding the upper third of each plant. Colors include blue, pink, red, and white. Use larkspurs in mixed borders or plant them in cutting gardens; their long stems and pastel colors make them graceful material for arrangements.

### *Cosmos bipinnatus*
Cosmos

Native to Mexico, where it thrives in poor soil and high heat, cosmos is almost ridiculously easy to grow. *C. bipinnatus,* the most popular cosmos species, displays 4-inch, daisylike flowers in white, pink, red, or purple, enhanced by delicate leaves on long stems.

After all danger of frost has passed, scatter seeds onto the soil, water them, and watch them sprout within days and shoot to a height of up to 5 feet. To encourage bushiness, however, pinch the growing tip when plants are about 12 inches tall; plants allowed to grow too tall may become top-heavy and break. Use *C. bipinnatus* in borders as a backdrop for shorter plants. They are also good for cutting.

A less common easy-care cosmos, *C. sulphureus,* grows much more compactly and possesses even better heat tolerance. Its color range includes yellow and orange. Two dwarf compact varieties of *C. sulphureus,* 'Sunny Yellow' and 'Sunny Red' (actually orange-scarlet), stay below 2 feet, which makes them suitable for planting in beds.

### *Dahlia* **hybrids**
Dahlia

The easiest dahlias to grow are the dwarf bedding types produced from seed. They make colorful mounds, 1 to 2 feet high, with daisylike flowers up to 4 inches wide in a rich color assortment including white, yellow, orange, red, purple, and peach. Some bedding dahlias have bright green foliage, others have bronze foliage. The larger-flowered dahlias are grown from tubers, but these tend to be slow to bloom, and they need constant feeding, staking, and watering to keep them in flower.

*Eschscholzia californica*

*Gazania rigens* 'Sunshine'

Bedding dahlias are best purchased in six-packs if you don't want to fuss with starting them from seed, which takes six to eight weeks to produce a good transplant. Give dahlias sunny, humus-rich loam soil, and water them at the roots whenever the soil feels dry. Avoid drenching the leaves with water because it encourages powdery mildew, which discolors the plants with a gray mold. To keep leaves dry, use drip irrigation or a watering can with a long spout. Remove faded blooms to keep the plants flowering.

Cooler coastal gardens and fog belts foster the best displays, although dahlias will take heat if they are watered regularly. The award-winning, American-bred 'Redskin' variety (2 feet tall) has bronze leaves and an extensive range of flower colors. 'Rigoletto' (15 inches tall) has bright green foliage and a good assortment of flower colors.

## Eschscholzia californica
California-poppy

One of the most uplifting sights in all of nature is the California hills in springtime, clothed in native poppies whose shimmering orange petals reflect the sunlight like newly spun silk. California-poppies are easy to grow from seed sown directly into the garden, either in early spring for early summer flowering or in late summer for flowering early the next spring. They sprout thick and fast, blooming within 50 days in mounded clumps that grow to 15 inches high. They self-seed readily and are a popular component of wildflower mixtures.

California horticulturist Luther Burbank was largely responsible for extending the color range of California-poppies to white, yellow, pink, and red. Available now in mixtures, they like full sun and sandy soil that drains well. Plant them in mixed beds and borders, in rock gardens, as an edging for paths, or massed on sunny slopes. They are cool-season plants, flowering best

in spring, early summer, and fall. The silver-blue lacy foliage provides an attractive display when the flowers are out of bloom.

## Gazania rigens
Gazania

Native to harsh, semidesert areas of South Africa, gazanias are easy to grow in even poor, sandy soil. There are two types: clump-forming, which are used mainly as annuals throughout North America; and trailing, which are vigorous, spreading plants employed most often as perennial ground covers in California and the Southwest. Both enjoy full sun and produce an abundance of 4-inch, daisylike flowers that close on cloudy days.

Clump-forming gazanias produce compact, 12-inch-high rosettes of leaves that are usually dark green. The masses of 4-inch daisylike flowers bloom in spring or summer, depending on variety and location.

Trailing gazanias are mostly low, spreading plants used as ground covers in mild-winter areas such as southern California. About the same height as clump-forming gazanias, they have all-gray leaves and 2½-inch flowers that may bloom in winter as well as in spring and summer. Flowers of both types may be red, yellow, orange, white, or purple, most with attractive bronze or black centers.

Use annual, clump-forming gazanias to edge beds and borders, or wherever you need a summer-long flowering ground cover. Use perennial and trailing kinds as permanent ground covers in areas with mild winters. An excellent annual mixture of all the best clumping gazanias is Sunshine bicolors; it's easy to grow these from seed sown directly into the garden. 'Copper Canyon' is a popular winter-flowering, trailing perennial with striped orange-and-copper petals. It's meant for warm-winter areas and is best purchased ready grown from local garden cen-

*Gomphrena globosa* 'Strawberry Fields'

*Helianthus annuus*

ters. 'Sungold' is a spectacular evergreen trailing perennial with yellow flowers. By means of vigorous underground stems, it quickly spreads into a dense carpet of silvery blue leaves. Its salt tolerance makes it especially useful for coastal gardens. Both trailing and clumping gazanias are usually sold in six-packs propagated from cuttings.

### Gomphrena globosa
Globe-amaranth

When you travel through the southern states during severe drought and see the shriveled brown of flowering plants that have succumbed to heat and dryness, you'll also see the papery flower heads of globe-amaranth still cheerfully blooming away. Native to the parched plains of India, globe-amaranth is blessed with both

a cast-iron constitution and good looks. Colors include purple, orange, yellow, lemon, pink, and white.

Plants grow 2 to 3 feet high, creating a perfect mound. The globular flowers bloom profusely all summer until fall frost. The more flowers you cut for arrangements, the more new flower buds will appear. Plant globe-amaranth in sunny, well-drained beds, borders, or cutting gardens. A dwarf variety, 'Buddy' (6 inches tall), is deep purple—good for edging and pots. Taller varieties make an attractive temporary hedge at the back of a border.

For everlasting arrangements, dry the flowers by hanging them upside down in bunches. They retain their perky colors indefinitely. Seeds germinate easily, either sown directly into the garden or started indoors six weeks before transplanting. Six-packs of seedlings are not yet sold at many garden centers, but when available they are inexpensive.

### Helianthus annuus
Sunflower

In keeping with their ample size, sunflowers produce large, easy-to-handle seeds that germinate reliably in the garden after frost danger has passed and bloom in summer and fall.

Although the 2-foot heads of the giant varieties provide quite a spectacle at their peak (midsummer), the smaller-flowered kinds bloom more profusely on bushier plants, making them better ornamentals. Moreover, the small kinds reach their peak late in the season when other flowering plants have started to dwindle. Try the widely available Color Fashion seed mixture (6 feet high), which offers 6-inch sunflowers in a beautiful assortment of colors—yellow, orange, red, and bronze around dark centers. Some flowers incorporate all these colors in concentric bands.

Use the tall sunflowers as a screen at the back of a border. Harvest the flowers for

arrangements and the plants will keep producing more. In recent years, plant breeders have developed some interesting dwarf varieties, such as 'Teddy Bear' (3 to 4 feet tall), whose 6-inch, double, golden blooms call to mind the famous still life by Vincent van Gogh. 'Mammoth Russian' is a tall variety to grow if you want huge, single yellow blooms that become loaded with nutty seeds relished by both songbirds and people.

### Helichrysum bracteatum
Strawflower

One of the world's great wildflower displays occurs in October (springtime) on the western coast of Australia, where entire hillsides are transformed into colorful seas of native strawflowers. In North America strawflowers have become the favorite "everlasting" flower for floral arrangers, since their color range is the most extensive and most intense among flowers suitable for drying. The

*Helichrysum bracteatum*

*Impatiens wallerana*

*Ipomoea tricolor* 'Heavenly Blue'

daisylike flowers, up to 3 inches across, have pointed petals with a papery texture and a satin sheen. Colors include yellow, white, orange, and pink.

Strawflowers grow as easily as dandelions. Sow seed directly into the garden after frost, or purchase inexpensive six-packs from local nurseries. In full sun and well-drained loam soil, they bloom within eight weeks of sowing seed and continue all summer until severe fall frosts.

Plants grow 1½ to 4 feet tall, depending on variety. 'Bright Bikini' is a dwarf mixture with a mounded form suitable for bedding. Since stems of all strawflowers wilt fast and must be replaced by wires in dried arrangements, the short stems of 'Bright Bikini' are no handicap. Monstrosum, the largest-flowered mixture (3 inches across), comes in an assortment of bold colors and grows to 4 feet and more, making it an excellent background for mixed flower borders.

## Impatiens wallerana
Impatiens, patienceplant

No flowering plant is more popular for a shady location than impatiens. Not only does it have an extensive range of colors (15 at last count), but it flowers nonstop, gathering momentum soon after transplanting and continuing lavishly all summer. The mound-shaped plants are classified as dwarf (1 foot), medium (2 feet), or tall (3 feet or more), and spread to several times their mature height. The flowers are small (up to 2 inches across) but numerous; color choices include white, crimson, scarlet, pink, salmon, orchid, rose, and orange, and a range of bicolors. Some varieties have double flowers.

Although famous for its vibrant, prolonged displays in shade, impatiens will also grow in full sun provided that the soil is cool and moist. To

achieve these conditions add plenty of humus (especially peat, compost, or leaf mold), water frequently, or surround the beds with brick or flagstone to insulate the soil.

For the best flowering display, choose hybrids. The 'Super Elfins' and 'Futuras' are early, larger-flowered (2 inches) than older types, and dwarf (8 to 12 inches high). They are thus especially suitable for tubs, window boxes, and hanging baskets. For even larger blooms (2½ inches) and a taller habit (2 feet), grow Blitz hybrids, especially the award-winning 'Red Blitz.'

Growing impatiens from seed is not easy. The seed is small and needs pampering for 10 weeks to reach transplanting size. Seedlings are also highly susceptible to damping-off disease, a fungus that rots the base of the stem. Instead of bothering with seed, buy young plants from a garden center. You'll usually find them with a few flower buds opening, so that you can pick out the exact colors you desire.

## Ipomoea tricolor
Morning glory

No flowering annual will cover an ugly fence or bare wall faster than morning glories, nor produce such a spectacular show of flowers. All morning glories demand is a trellis or some other form of support.

Despite the extensive range of newer varieties, which includes white, red, pink, blue, and bicolored flowers, the traditional 'Heavenly Blue' variety remains the all-time favorite. A remarkable vine that can grow 12 feet high in a summer, it has trumpet flowers that flare out to 4 inches across. These usually close by midafternoon, however, except on cloudy days.

The easiest way to grow morning glories is to first soak the hard seeds overnight in lukewarm water. By morning, the seed coats will have softened and the seeds will have soaked up enough moisture to ensure almost instant germination when placed in the

Lavatera trimestris

Lobularia maritima

Nicotiana alata 'Nicki-Red'

garden. Plant in full sun in well-drained loam soil after the last spring frost, placing seeds 1 inch below ground and 3 inches apart. Morning glories don't mind crowding; they will "knit" together to produce a dense vine cover. Overly fertile soil may result in too many leaves at the expense of flowers, but the heart-shaped leaves are decorative in their own right.

### Lavatera trimestris
Mallow

A Mediterranean native perfectly suited to most North American conditions, mallow tolerates both prolonged heat and cool coastal summer fog. It grows quickly from seed sown directly into the garden after frost, flowering within 75 days. The 4-inch-wide flowers, which resemble hollyhocks, are so profuse at midsummer that they may completely hide the dark green foliage.

Mallow is not particular about soil, provided that it is sunny, drains well, and receives water during dry spells. Plants grow to 3 feet high, in a mounding form, and are best used for background in mixed flower borders. They also make an attractive, summer-flowering temporary hedge. The flowers are dazzling, shimmering like satin in the sunlight. Colors include white, pale pink, and dark pink. The award-winning 'Loveliness' variety is a beautiful, deep rose pink.

### Lobularia maritima
Sweet alyssum

This plant doesn't make a big statement—it simply fills in between the other flowers, lending a delicate finishing touch to the garden. Plants grow 6 inches high, creating a spreading mound of ¼-inch white, pink, or violet blue flowers that often completely hide the fine, needlelike foliage. The tiny flowers have a honey scent.

Use sweet alyssum in mixed borders as a low edging that will spill over into pathways for a pleasantly informal look. Use it also to create a temporary ground cover, to fill cracks between flagstones and in dry walls, to create drifts in a rock garden, or to cascade over the rims of window boxes and hanging baskets.

Alyssum is simple to grow: Just sow seeds where you want them to bloom. They will germinate like weeds within a week and start flowering within 40 days. There is no need to thin them, since they tolerate crowding. They'll even self-sow to return the following year. About the only soil condition they dislike is waterlogging.

The most striking varieties are the award-winning 'Carpet of Snow' (pure white) and 'Wonderland' (deep rosy pink). Although all alyssums are easy to grow from seed, local garden centers usually offer a good selection in six-packs.

### Nicotiana alata
Nicotiana, flowering tobacco

Older varieties of nicotiana close up in the afternoon, but newer hybrids not only stay open but flower more profusely over a longer period, tolerating both high heat and cool conditions. If flowering starts to diminish by the end of summer, simply cut down the flowering stems to just above the soil and water them conscientiously until new flowering stems appear, usually within a few weeks. The second blooming cycle lasts several weeks.

Related to petunias, nicotianas are tender annuals best purchased as seedlings from local nurseries. The rosette of broad, crinkled, dark green leaves sends up a long, slender flower stalk to 4 feet high, topped with a cluster of star-shaped, tubular flowers up to 2 inches across. The color range includes red, pink, white, yellow, purple, and a particularly appealing lime green. Plant in full sun or partial shade, in well-drained loam soil. Water

*Papaver rhoeas*

*Pelargonium* × *hortorum*

regularly to keep the soil moist. Use tall (4-foot) nicotianas for backgrounds; dwarf (2-foot) varieties massed in beds, borders, or in tubs on decks and patios. The dwarf 'Nicki' series of hybrids (1½ to 2 feet) includes the award-winning 'Nicki-Red'.

### *Papaver rhoeas*
Shirley poppy

Like most other poppies, Shirley poppies hate to be transplanted, since the least root disturbance makes them wilt and die. However, seed scattered on sunny, well-drained soil will germinate easily within a week and come to full bloom within 75 days. Plants grow 2 to 3 feet high and produce 4-inch flowers that have prominent black centers and, sometimes, attractive black petal markings. Colors include red, white, pink, and bicolors. Mass Shirley poppies in beds and borders or include them in meadow gardens, where they

look especially fine with bachelor's-button (*Centaurea cyanus*).

Shirley poppies will not bloom through hot, dry summers, so sow the seed in early fall—even several weeks before frost—to give the plants a long blooming period before summer arrives. Sowings made in September will produce stocky young plants that can survive freezing winters to bloom extra early the following spring. In cool-summer climates and coastal fog belts, Shirley poppies will bloom all summer from repeated sowings at two-week intervals.

### *Pelargonium* × *hortorum*
Geranium

These versatile plants have such dependable constitutions that it is little wonder nurseries sell more of them than any other flowering annual. Use them massed in beds and borders, alone in pots, or lined up in window boxes. Simply give them full sun and humus-rich well-drained soil, keep the soil moist, and remove spent blooms, and they will bloom from early summer until fall frost.

Umbrellas of small florets form the attractive 4- to 5-inch flower clusters, held erect by slender stems. The plants are mound-shaped and grow 1 to 2 feet tall, depending on variety. The 'Ringo' series contains a wide color range, including scarlet, rose, pink, salmon, white, magenta, orange, and bicolors. Some geraniums have a dark brown band, called a horseshoe, around their scalloped green leaves.

Geraniums can be grown from seed or purchased as seedlings in six-packs. Six-packs grown from seed are less expensive than those grown from cuttings and usually produce a longer-lasting display. All geranium flowers tend to shatter after hard rainfall.

### *Petunia* hybrids
Petunia

Most petunia hybrids are classified as either grandiflora (large-flowered) or multiflora (many-flowered). People who see them flowering side by side at a garden center usually choose the grandifloras—a mistake! Although multifloras are just half the size of grandifloras (blooms are 2 to 3 inches, rather than 5 inches, wide), there are so many more blooms per plant that they put on a far better show. They also recover more quickly from rainstorm damage and, more importantly, are easier to grow.

Although petunias have a long flowering period (5 or 6 weeks), they may exhaust themselves by midsummer and will die unless you give them a "haircut." Trim the plants almost to soil level, keep them watered, and in late summer they will sprout new leaves

*Petunia* hybrid 'Summer Madness'

*Portulaca grandiflora* 'Afternoon Delight'

*Rudbeckia hirta* 'Goldilocks'

and flower buds to produce another long-lasting flush of color that may continue until fall frost.

Petunias can be propagated from seed sown indoors eight weeks before transplanting into the garden, although flats and six-packs can be purchased inexpensively at garden centers. Plant them in full sun in humus-rich, well-drained soil. Petunias thrive in all well-drained soils in all states, even Alaska.

The compact, mounding plants usually grow 12 inches high. When in bloom, they are covered with trumpet flowers, either smooth-petaled (multifloras) or ruffled (grandifloras). Color range includes white, yellow, red, pink, blue, purple, and bicolors. Some petunia varieties are single, some are double. Mass them in low beds and borders, or plant them in tubs, window boxes, or hanging baskets.

The Joy series of multifloras is extremely vigorous and free flowering. The new Ultra series of grandifloras has a "basal-branching" growth habit—plants branch strongly from the base, forming a low, spreading, bushy profile. They are thus more resistant to wind and rain than other grandifloras.

## Portulaca grandiflora
Portulaca

If you have a hot, dry, sunny spot where it is difficult to get anything to grow, portulaca may love it! A low, ground-hugging, succulent from the semidesert regions of South America, portulaca blooms within just 40 days of sowing seeds directly into the garden. (Be sure to wait until frost danger has passed.) Plants spread 3 feet or more, yet rarely exceed 6 inches in height. They tolerate crowding and will mingle leaves and stems in a dense ground cover.

Portulaca covers itself with cupped single or semi-double flowers resembling miniature roses. Colors include white, yellow, orange, magenta, pink, red, salmon, and purple; petals have an attractive sheen. Choose hybrids such as the Afternoon Delight series for a high percentage of fully double blooms measuring up to 2 inches across—almost twice the size of nonhybrids. These flowers do not close up in late afternoon as do other varieties.

Use portulaca massed in low beds and borders. A related species, *P. oleracea,* has smaller, single flowers and is a good choice for hanging baskets and window boxes because of its low, spreading growth habit and high tolerance of heat. It tolerates even greater heat stress than the larger-flowered types.

## Rudbeckia hirta
Black-eyed-susan, gloriosa daisy

Native to the waysides of North America, *Rudbeckia hirta* is a valuable summer-flowering plant, displaying golden yellow, daisy flowers with chocolate brown centers. It relishes the heat, tolerates poor soil, and prefers full sun or light shade.

Plant breeders have created fancy cultivars, called gloriosa daisies, which can be treated as annuals because they bloom the first year from seed sown directly into the garden. However, as perennials, they live for many years. Seed germinates so reliably that you can sow it on top of frost or snow; it slips into the soil as the snow melts and germinates as soon as the ground begins to warm in

*Salvia farinacea*

*Salvia splendens*

spring. You can also start seed indoors. Much larger-flowered than the wild types, gloriosa daisies grow flowers up to 5 inches across on bushy 3-foot-high plants. The range of flower forms and colors is also more extensive, including single, semidouble, and double varieties in mahogany, bronze, and two-color combinations. Use them in mass plantings, borders, and beds, and as cut flowers.

A related species, *R. fulgida*, includes the perennial variety 'Goldsturm', which grows 2 feet tall and produces so many flowers that they almost completely hide the dark green, spear-shaped leaves. Although these flowers are smaller than those of gloriosa daisies (merely 3 inches across), they are much more profuse; and the leaves are resistant to powdery mildew, a filmy white fungus that attacks other kinds of rudbeckias. Look for 'Goldsturm' in containers, since it can be propagated only from cuttings.

## Salvia farinacea
### Blue salvia

A native North American wildflower from the dusty plains of Texas, blue salvia is a true survivor, growing in every state and tolerating both high heat and cool conditions. Although grown mostly as an annual, it is a perennial in southern states and in California. Its mass of elegant, violet blue flower spikes is held erect above 3-foot-tall upright plants with slender, silvery gray leaves. There is also a white variety, popular in gray gardens and all-white gardens.

Use blue salvia as a background in mixed borders or as a highlight in island beds. It's also valuable as a cut flower and as a tall highlight in container plantings. Purchase seedlings and transplant them into sunny, well-drained, humus-rich loam soil after frost danger has passed. Snip off the first flower bud to

encourage bushier growth. By far the most attractive variety is 'Victoria', which has the deepest blue color and a natural tendency to branch low from the base.

## Salvia splendens
### Scarlet sage

Few, if any, flowering plants can match the brilliant red of scarlet sage. Its intense coloring is especially effective against a background of evergreens or in island beds surrounded by grass or edged with boxwood.

Plants grow 1 to 2½ feet tall, depending on variety. The tubular flowers are arranged on bold spikes held above decorative, blue-green heart-shaped leaves. Although crimson is the most popular color, plant sellers now offer white, purple, and pink.

Since seeds take at least eight weeks to grow to transplantable size, it's better to purchase these tender plants in flats or six-packs from the nursery for setting outdoors after frost. In northern gardens, plant them in full sun; in southern gardens, light

shade. Plant in well-drained, humus-rich soil and remove faded flowers to ensure nonstop flowering until fall frost. To encourage a bushier plant, snip off the first flower spike after transplanting. During hot and dry spells, keep the soil moist through regular watering.

Compact dwarf varieties such as the award-winning 'Carabiniere' (1 foot) are suitable for low beds and borders. 'America' (2½ feet), the tallest scarlet sage, is best as a background or used sparingly as a highlight in mixed flower borders. All kinds make dramatic container plantings.

## Tagetes patula
### French marigold

Planted in full sun, French marigolds are practically indestructible. They are easy to start from seed sown directly into the garden, are highly resistant to pests and diseases, and need little watering or fertilizer. Their color range is

*Tagetes patula* 'Janie'

*Thunbergia alata*

*Tropaeolum majus*

dazzling—clear yellow, glowing orange, burnished bronze, and lively two-tone combinations. Unlike its taller cousin the American marigold (*T. erecta*), the tidy French type blooms over a long period and never grows spindly. It's a vibrant contributor to beds, borders, edgings, and container plantings, as well as to floral arrangements.

Sow seeds into well-drained loam soil after the last spring frost. In poor soil add low-nitrogen fertilizer only, because too much nitrogen stimulates leafy growth at the expense of flowers. In good soil, there is no need to fertilize at all. Keep soil moist until seeds germinate; then thin seedlings to 12 inches apart. Once established, plants will tolerate short dry spells, but plan to water once a week unless there is rain.

For truly abundant flowering, try the new triploid hybrids, a cross between the dwarf French marigold and the tall American. Also called mule marigolds, they are unable to produce viable seed

and instead pour all their energy into nonstop blooming. From early summer to fall frost, these hybrids create a greater density of color over a longer period than any other kind of marigold. Buy triploids ready grown in six-packs, since seed germination can be erratic. 'Red Seven Star' has more reliable seed than the others, and its scarlet and gold 3-inch flowers are amazingly prolific on the 18-inch-high bushy plants.

Among standard varieties, consider 'Queen Sophia', an award winner with broad, showy flower heads and a rich russet-red coloring overlaid on golden yellow. Just 12 inches high and perfect for bedding or edging, it flowers as vigorously as a hybrid but is much less expensive.

In spring, local garden centers offer many varieties of marigolds. Some of these will already be starting to bloom so you can choose exactly the colors you want.

## Thunbergia alata
Black-eyed-susan vine

The growth rate of black-eyed-susan vine is extraordinary—it will grow more in one week than other vines grow in a

year. Within 60 days of transplanting, this unusual vine starts to produce bright orange flowers some 3 inches across. In full sun (or partial shade) and moist, humus-rich loam soil, it grows as much as 6 feet in one week, producing thousands of flowers until fall frost. Native to South Africa, black-eyed-susan vine tolerates high heat but grows and blooms best when nights are cool.

The heart-shaped bright green leaves resemble those of morning glory and, like morning glory, this plant is perfect for covering an ugly chain-link fence or decorating a wall, arbor, or trellis. Black-eyed-susan vines also make excellent plants for hanging baskets, their trailing stems cascading downward several feet. Plant three to a 12-inch pot for a dense, cascading effect. The cheerful flowers attract both hummingbirds and butterflies.

Sow seeds directly into the garden, or buy plants in containers at the nursery. Orange is the most popular color, but there are also white and yellow-flowered varieties, all with black eyes.

## Tropaeolum majus
Nasturtium

The pea-size seeds of nasturtiums are easy to handle and can be sown directly into the garden after spring frost. Bedding varieties such as the newer Whirlybird series (12 inches high) grow quickly when nights are cool and flower within 60 days. In addition to dwarf, mounded types like Whirlybird, there are vining kinds that grow to 6 feet high. Only prolonged summer heat and overly fertile soil will inhibit their flowering. Plants will recover from the former as soon as cool fall nights arrive, and will flower nonstop until the frost first.

The 2½-inch, flat-faced flowers of classic nasturtiums have spurs containing nectar that is attractive to hummingbirds, but which force the flowers to face the ground. In Whirlybird varieties these spurs are absent, and the flowers face the sky, making a more colorful display. Whirlybird colors include

*Verbena*

*Zinnia elegans* 'Torch'

red, yellow, white, pink, orange, and mahogany, plus bicolors.

Plant nasturtiums in full sun in unfertilized soil, and water whenever the soil feels dry. Use the dwarf varieties for bedding, edging, and container plantings, or choose the taller vining types to cover trellises and chain-link fences or to trail over slopes.

All parts of the nasturtium except the roots are edible. Flowers and leaves are added to salads to impart a pungent, slightly mustardy flavor. The seeds are often pickled for sale as "capers."

### *Verbena* hybrids
Verbena

Like many flowering annuals bred from desert species, verbenas can tolerate daytime heat as long as the nights are cool. Most verbena hybrids are low-growing, spreading plants with dark green serrated leaves and flowers arranged in clusters of primrose-like florets, up to 2½ inches across.

Colors include white, blue, red, and pink, some with a contrasting white or red eye.

Use verbenas as a low ground cover in beds and borders, grouping colors separately or in mixtures. Also use them in window boxes or hanging baskets, where their long, pliable stems create a cascading effect.

Since seed is difficult to germinate, leave it to the experts and buy ready-grown transplants from local nurseries. Plant them into sunny loam soil with good drainage, and water regularly during dry spells to maintain continuous bloom. Outstanding varieties include the award-winning 'Amethyst' (blue with white eye) and 'Blaze' (crimson with white eye).

### *Zinnia elegans*
Zinnia

Sow zinnia seeds into bare, sunny soil after frost and they will come up like weeds, sprouting within five days and flowering within six weeks. The best choices are the so-called cut-and-come-again

varieties, which flower more abundantly the more you cut them, providing a nonstop floral display until fall. The rounded, broad-petaled flowers of these zinnias are not the largest (2½ inches compared with some hybrids), but what they lack in size they more than make up for in quantity of bloom—up to three times as many as other types. Their long, 3-foot stems make zinnias good subjects for flower arrangements.

Most zinnias are classified as dahlia-flowered (with daisy petals, including the cut-and-come-again types) or as cactus-flowered (with pointed petals). Flowers of some hybrids, such as the dahlia-flowered 'State Fair' and cactus-flowered 'Zenith', grow to 7 inches across. The color range of zinnias is the most extensive among annuals, including many shades of yellow, orange, red, pink, purple, white, and two-tone combinations. There's even a grass green variety, 'Envy'.

Native to semidesert areas of Mexico, zinnias relish the heat and dislike cool, wet summers and coastal fog belts. In

these damp conditions they may succumb to powdery mildew, a gray mold that covers leaf surfaces. Overhead watering also promotes the disease. If mildew is a problem try the Ruffles hybrids—dahlia-flowered, cut-and-come-again types that are almost immune.

Zinnias purchased ready grown for transplanting may be sensitive to transplant shock unless the rootball is handled carefully and planted intact. When buying young zinnias, avoid plants with flowers showing; these may not survive transplanting. During the blooming period, snip off faded blooms to prevent seed formation, which drains the plants' energy and shortens the flowering display.

Use dwarf series, such as 'Peter Pan' (12 inches high, a dahlia-flowered type), for low, mass displays or container plantings. Plant taller kinds for a hedge effect or back-of-the-border display. Because of their vibrant colors, zinnias are effective planted in mixtures.

*Achillea filipendulina*
'Coronation Gold'

*Agapanthus africanus*

*Alcea rosea*

## PERENNIALS

Perennials add seasonal color to the garden. Their long-lived roots enable many of them to flower year after year although some, called biennials, live just two years. Many perennials grow easily from seeds or bulbs but do not bloom until the second year. For faster flowering, buy plants in containers.

### Achillea filipendulina
Yarrow

It's almost impossible not to succeed with yarrow. Even in poor soils and prolonged drought, this hardy perennial spreads and flowers profusely, crowding out even the most noxious weeds. The fernlike and fragrant dark green leaves are prized by herbalists for use in potpourris. The flat, round flower clusters, made up of many tiny yellow flowers, are held erect on long stems, which can be cut to make long-lasting dried arrangements that retain their color and aroma for years. Plants grow to 4 feet high in full sun; they bloom in late spring through midsummer.

Use plants sparingly in mixed perennial borders and herb gardens; too many can be monotonous. The hybrid 'Coronation Gold' is more compact (3 feet) and more heat resistant than the species. *A. millefolium,* a close relative of *A. filipendulina,* is almost identical to it in appearance and just as aggressive, but flowers in white and shades of pink and red. 'Fire King', with rosy red flowers and silvery foliage, is outstanding for its bright color.

Purchase plants in containers in the spring. Divide clumps every second or third year to prevent them from crowding other desirable perennials.

### Agapanthus africanus
Lily-of-the-Nile

If you happen to live in the South or California, this spectacular flowering bulb is well worth cultivating. Persistence in poor soil and drought make lily-of-the-Nile a favorite for areas that are subject to neglect, such as parking strips. It is also popular in seashore gardens. Its elegant, rounded

blue or white flower heads—up to 10 inches across in some hybrids—are held erect on poker-straight, 3- to 4-foot stalks above strap-shaped, arching leaves.

Although lily-of-the-Nile will tolerate some shade, it prefers a sunny, well-drained site. Flowering usually begins in early summer and continues into autumn. In colder climates, these tender plants can be grown in pots and taken indoors for the winter.

Start with container plants from local nurseries or create new colonies by replanting pieces of rhizome (swollen underground stems) from any well-established clump. Headbourne hybrids, the hardiest varieties, are reportedly hardy as far north as Washington, D.C., on the East Coast; the flowers are deep blue or pale blue. 'Peter Pan' is a dwarf variety excellent for mass plantings to create a ground-cover effect.

### Alcea rosea
Hollyhock

This plant is admired for its tall stature and year-after-year persistence in the garden. Although there now are annual kinds and dwarf kinds with double pom-pom flowers, it is the old-fashioned tall perennial hollyhocks, with their ruffled, cupped flowers, that remain the most widely planted. They are easy to grow, requiring no special attention, and they thrive in sun and heat. Even poor soil will suit hollyhocks as long as it is sunny and well drained. Colors include red, white, pink, and yellow.

Seed of annual kinds may be sown directly into the garden after frost danger has passed, to bloom in midsummer. Thin the seedlings so that final plants stand at least 12 inches apart. Perennial kinds may be purchased in containers from local nurseries.

Use the tall types (up to 7 feet) at the back of flower borders or in groups as vertical accents. They look especially attractive towering above picket fences and low hedges.

*Aquilegia* hybrid 'McKana Giants'

*Armeria maritima*

*Asclepias tuberosa*

Summer Carnival is a stunning mixture of tall hollyhocks, displaying double pom-pom blooms up to 5 inches across. The award-winning Majorette mixture produces mound-shaped, low-growing plants with semidouble pom-pom blooms up to 3 inches across, in a wide color range, suitable for beds and borders.

### Aquilegia hybrids
Columbine

Developed from species native to North America and Japan, hybrid columbines are mostly short-lived perennials, although some varieties bloom in the first year from seed, like annuals. The gray-green foliage forms lacy mounds with a mass of pendant, tubular flowers held erect on long slender stems. Conspicuous spurs sweep back from the flower head, creating an intricate flower that is also known as granny's bonnet. Color range includes white, blue, pink, yellow, red, and bicolors.

Plants grow 3 feet high and prefer sun or partial shade and a cool, moist, humus-rich well-drained soil. Flowering occurs in late spring. Seed of 'McKana Giants', started eight weeks before last frost and transplanted outdoors, will produce flowers the first year. If you don't want to start from seed, buy transplants from local nurseries.

Use columbines in beds and borders, in rock gardens, or naturalized at the edge of woodland. They are also exquisite in arrangements.

### Armeria maritima
Thrift

One of Europe's more delightful sights is that of thrift cheerfully blooming at the edges of sea cliffs, seemingly impervious to gale-force winds, blinding sun, and destructive salt spray. Although few home gardens ever subject them to such abuse, these hardy perennials are up for almost any kind of neglect. They thrive even in poor, dry, sandy soil—indeed, well-watered and overly fertile conditions may foster root rot.

For several weeks in early summer, a lovely profusion of globular pink 1-inch flowers bloom on slender, erect stems about 8 inches above the foliage. The low cushion of blue-green, grassy evergreen leaves remains attractive even after flowers are spent.

Use thrift to cover sunny slopes, as an edging for paths or perennial borders, or massed in rock gardens. Its seed heads make attractive dried flowers. Divide well-established clumps to make new plants, or purchase container plants in spring.

### Asclepias tuberosa
Butterfly weed

Well deserving of the term *weed* because of its vigorous growth, butterfly weed is one of the most eye-catching perennials for full sun. Flaunting masses of brilliant orange,

3- to 4-inch flower clusters, it attracts clouds of butterflies to the garden in early summer. It thrives even in dry, infertile soil, and competes well with meadow grasses. Hardy into Canada, it is also among the most heat- and drought-tolerant perennials. Its only dislike is soil that is constantly soggy and thus prone to rot the long, fleshy taproot.

Plants grow to 3 feet high and form erect clumps that spread to several times their height. Use butterfly weed in mixed perennial borders, massed in a rock garden, or naturalized in meadow gardens; flowers are popular in arrangements despite the sticky, milky sap the stems exude when cut. In addition to orange, there is a yellow form and a seed mixture, Gay Butterflies, that includes red and pink. Although plants are easy and inexpensive to raise from seed sown into sandy soil outdoors, they are more convenient to buy in containers locally.

Aster novae-angliae 'Superbus'

Astilbe

Aurinia saxatilis

### Aster novae-angliae
New England aster

These colorful flowers are so essential a part of fall gardens that it is hard to envision a fall perennial border without them. Native to the eastern United States, they are the hardiest cultivated asters, producing numerous 2-inch, daisy flowers on billowing plants that range from 1 to 6 feet in height, depending on variety. Colors include white, pink, blue, and purple.

Plants prefer sunny, well-drained loam soil and weekly watering during dry spells. The 'Alma Potscke' variety is a superior hybrid, remaining reasonably compact at 3 feet and with flowers the closest to red of any New England aster. Plants bought in containers at garden centers generally benefit from shearing of the tips to encourage side branching and hence compact, bushy growth. Plants spread rapidly in fertile soil and may need dividing every other year to keep them healthy. The taller varieties are excellent for cutting; a single flowering stem makes an instant bouquet.

Use New England asters in mixed beds and borders where autumn color is needed. Also try them along stream banks and pond margins, in wild-flower meadows, and at the edge of woodland. They are good companions to yellow fall-blooming sunflowers.

### Astilbe × arendsii
Astilbe, false-spirea

If you have a lightly shaded area with a cool, moist soil that can be enriched with humus (in the form of peat or leaf mold), by all means try astilbes. Their brilliant, feathery flowers may well outshine everything else in the garden. The dark green ferny foliage is decorative even when the flowers are absent.

Favorite planting sites are at the edge of a lawn, along the bank of a stream or pond, or in a rock garden. They make excellent companions to

hostas and ferns. Use them also in flower arrangements. Buy container plants from local nurseries or increase established plants by division.

Many of the best astilbe hybrids were developed in Germany, including 'Deutsch-land' (white), 'Fanal' (carmine red), and 'Europa' (pale pink). A related species, *A. chinensis* (especially pink 'Pumila'), is more tolerant of full sun and drier soil. It is lower growing (1½ feet) and spreads by underground runners, making a good edging for perennial beds and borders.

### Aurinia saxatilis
Basket-of-gold, yellow alyssum

In gardens where flowers must produce "a big bang for the buck" or else be relegated to the compost pile, basket-of-gold is bounty indeed. In any sunny, well-drained garden soil it flaunts a super-abundance of bright yellow flower clusters. Although each individual four-petaled floret is small (about ¼ inch), their combined effect is that of a spreading cushion of flowers. These appear in early spring in 8- to 12-inch-high spreading

mounds that cascade over borders and walls. Even when the plant is not in flower the gray-green leaves are attractive, although it is good practice to shear them to within several inches of the soil after flowering, to maintain compact, vigorous growth.

Plants grow easily from seed, or from container plants purchased at local nurseries. They tolerate drought and demand little or no fertilizing.

Use basket-of-gold as an edging for borders and beds, particularly those with a raised edge so that plants can cascade over the side; also plant it in cracks between flagstone paving and in dry walls. The 'Compacta' and 'Plena' varieties are a gorgeous buttercup yellow; 'Atrina' is paler lemony yellow. Basket-of-gold combines especially well with blue spring flowers such as forget-me-not and blue phlox.

*Baptisia australis*

*Chrysanthemum × superbum*

*Coreopsis lanceolata*

## Baptisia australis
False-indigo

*Australis* in Latin means "southern," and appropriately false-indigo is a native of the southern United States. Virtually indestructible and hardy into Canada, this lupinelike plant with indigo blue flower spikes grows to 4 feet high, forming clumps of erect stems.

Plants prefer full sun and are easily grown in any well-drained soil, including stony soil. They flower in spring. Remove faded flower stems to prolong flowering and to prevent formation of the plant's conspicuous black seedpods.

Use false-indigo as a highlight in rock gardens, herb gardens, or mixed beds and borders—especially as a background plant. Purchase plants in containers from a nursery, as seed takes two to three years to produce flowers.

## Chrysanthemum × superbum
Shasta daisy

Many kinds of annuals and perennials are popularly called daisies, a name for flowers with button centers and rays of evenly spaced, narrow pointed petals. Introduced by California horticulturist Luther Burbank, Shasta daisy has become one of North America's favorite perennials, valuable for garden display as well as for floral arrangements.

Shasta daisy is the largest-flowered of a group of gleaming white daisies with bright yellow centers. In addition to the usual single-flowered types there are double-flowered Shastas and Shastas whose flowers measure up to 6 inches across. Shasta daisies flower in midsummer and are most easily grown from container-ized transplants.

Closely related to Shasta daisies (all are types of chrysanthemums) are fever-few, oxeye-daisy, and nippon daisy, all of which are even more foolproof than Shastas. What they lack in flower size they make up for in sheer quantity of blooms.

Feverfew (*C. parthenium*) grows 1 to 3 feet tall depending on variety and has clouds of ½-inch white or yellow flowers. Oxeye-daisy (*C. leukanthemum*) is a slightly smaller version of the Shasta daisy, but flowers earlier; it grows wild along waysides throughout North America. The 2-foot-tall nippon daisy (*C. nipponicum*) forms a cushion of 3-inch flowers that bloom in fall. Include them all in the garden for color continuity from early summer to fall frost.

All these daisies prefer moist, well-drained, reasonably fertile soil and full sun (in cool climates) or light shade (where summers are hot). Pinching out the main stem after transplanting encourages bushiness and compactness. They are superb for mixed perennial borders and cutting gardens. Dwarf varieties of Shastas, such as 'Miss Muffet' (12 inches), will bloom from seed the first year and are good for low bedding. 'Starburst' (3 feet) is a tall type that produces extralarge flowers up to 6 inches across.

## Coreopsis lanceolata
Tickseed

Hardy and drought tolerant, tickseed seems to laugh at cold winters and smile at summer heat. A native of the North American prairie and southern states, this yellow daisylike perennial has rectangular black seeds that stick to clothing like ticks; hence its name. The extremely free-flowering plants grow to 3 feet high, forming bushy clumps; they tolerate poor, sandy soil.

Tickseed adds a bright highlight to mixed beds and borders. It thrives on sunny slopes and in wildflower meadows, where it reseeds itself freely. The long stems are good for cutting. Remove faded flowers to ensure continuous blooming.

In addition to tall varieties such as 'Sunburst', a double-flowered yellow, there are several dwarf types including 'Goldfink' (1 foot high). The award-winning 'Early Sunrise'

Cortaderia selloana

Crocus vernus

(2 feet) produces semidouble blooms in just 11 weeks from seed started indoors. For even faster blooming, buy container plants from local nurseries.

An equally carefree relative is threadleaf coreopsis (*C. verticillata*), a native of the eastern United States. Plants form 2-foot-high spreading clumps with feathery foliage and 2-inch, star-shaped yellow flowers. 'Moonbeam' is a particularly attractive cultivar with pale, primrose-yellow flowers.

### Cortaderia selloana
Pampas grass

Hardy into Canada on the West Coast, though not reliably hardy north of Washington, D.C., on the East Coast, pampas grass is a magnificent clump-forming ornamental that withstands heat and drought. Its silky white, 10-foot-high flower plumes create a dramatic lawn highlight and are valued for both fresh and dried arrangements.

Where pampas grass, which is a South American native, will not survive freezing winters, the hardier maiden grass (*Miscanthus sinensis* 'Gracillimus') can take its place; it has arching, smaller plumes and a height about half that of pampas grass.

Pampas grass is best purchased in containers. Plant it in spring into any sunny soil—wet or dry, acid or alkaline. It will grow quickly to form a fountain of slender, arching leaves with sawtooth edges that can cut fingers like a razor if touched. The plumes appear in late summer and persist well into winter.

These plants need room to spread. Space them at least 8 feet apart to make a windbreak, screen, or high hedge. If left alone, they will naturalize in the garden. Alas for dwellers in the Northeast and Midwest, where these otherwise ideal plants cannot survive.

### Crocus vernus
Giant crocus

What a joy and surprise crocuses are when they pop through bare soil—seemingly overnight—before the last

snow has melted, cheerfully announcing the end of winter! Crocuses can be easily naturalized in a lawn to reappear spontaneously each spring. In fall, simply peel back a 1- to 2-foot section of the turf, loosen the soil to a depth of at least 6 inches, and cover the corms (bulblike modified stems) with 4 inches of soil; then roll back the turf. The site should be well drained and in full sun. Naturalized plantings such as this must not be mowed until after the leaves have died, usually by the beginning of June, since the leaves help the corms rejuvenate themselves.

Plant crocuses in clumps of all one color or in drifts of mixed colors. In addition to the popular purple crocus, there are white, yellow, and bicolored varieties. Native to high-elevation areas of the Mediterranean, crocuses thrive in areas with cold winters. To encourage them to come back each year, feed with a high-phosphorus fertilizer in fall.

In spring, supermarkets offer crocuses in decorative containers, such as ceramic Dutch clogs, for extra-early flowering indoors. After the flowers have faded, the corms can be planted outdoors to rebloom the next year.

### Dianthus plumarius
Cottage-pinks

From colonial times through the 19th century, cottage-pinks were the most popular perennials in North American gardens—grown to remind nostalgic immigrants of cottage gardens in Europe. These fragrant hardy plants bloom profusely in full sun during the cool spring months, surviving hot summers if watered weekly during dry spells. A hybrid strain known as 'Alwoodii' is the most heat resistant.

The mound-shaped plants grow to 12 inches high, with dainty, fringed 1½-inch flowers in white, pink, red, and purple, many with contrasting eyes. The evergreen leaves are gray-green and grasslike.

*Dianthus plumarius*

*Dicentra eximia* 'Luxuriant'

*Digitalis purpurea*

Use cottage-pinks to edge beds and borders and to create drifts in rock gardens. Seeds germinate reliably within two weeks, preferring humus-rich alkaline or neutral soil; nurseries also sell plants in containers. They spread rapidly and are easily divided to make new plants.

If flowering diminishes during extended hot, dry spells, shear plants back to within 2 inches of the soil to force new growth and repeat blooming when cool weather returns in fall. In areas with cooler summers, such as coastal fog belts, plants will bloom all summer.

Closely related to perennial cottage-pinks are China-pinks (*D. chinensis*), among which there are many excellent annual varieties, particularly the 'Charms' series. These plants form neat mounds of color just 6 inches high and bloom prolifically during cool weather.

## *Dicentra eximia*
### Bleedingheart

Native to the eastern United States, *D. eximia* has been crossbred with *D. formosa,* a Pacific Coast native, to create hybrids that do well across the country. Although *D. eximia* is best suited to cool, moist soil in light shade, its hybrid offspring, such as 'Luxuriant', grow equally well in full sun and tolerate hot, dry climates. Avoid heavy clay soil and poor drainage, however. A high humus content works wonders.

The 1½-foot-high, spring-flowering plants create a decorative mound of finely cut, fernlike gray-green leaves with an abundance of heart-shaped, pendant pink flowers, borne in clusters on long stems. The cultivar 'Snowdrift' has white flowers.

Bleedinghearts are good for edging paths in woodland gardens; for massing as a ground cover, particularly in light shade; or as drifts in rock gardens. The long stems are suitable for cutting. Buy bleedinghearts in containers from local nurseries, since plants grown from seed take two years to flower. Plants in the garden will replenish themselves from seed, and well-established clumps can be divided to make new ones.

A related species, Japanese bleedingheart (*D. spectabilis*) grows shrubby to 3 feet high, with long arching branches and flower clusters that appear in spring. Although more dramatic than *D. eximia,* it flowers for only two weeks—disappointing when compared to the summer-long display of *D. eximia* hybrids.

## *Digitalis purpurea*
### Foxglove

There is hardly a place in the continental United States where foxglove will not grow, provided that plantings are timed so that flowering will occur during cool weather. The tall, stately flower spikes are studded with downward-facing tubular florets splashed in the throat with exotic spots. Colors include various shades of red, pink, purple, white, and yellow.

Plants prefer cool, moist, humus-rich soil in sun or partial shade. In areas of hot, alkaline soil, grow foxglove in pots or in raised beds filled with peaty potting soil.

Plants grow to 5 feet high, making striking cut flowers and good vertical accents at the back of a flower border. They will naturalize in woodlands and meadows. Foxglove is usually biennial, producing a rosette of tongue-shaped leaves the first year and flowering the next year in late spring or early summer. Each plant produces millions of tiny seeds that self-sow. The seedlings of the award-winning variety 'Foxy' will tolerate mild frost and freezing; they thus will flower the first year from seed started early indoors and transplanted several weeks before the last expected frost. If you prefer not to fuss with seed, seek out container-grown plants at the nursery and transfer them to the garden in spring or fall.

*Echinacea purpurea*

*Gaillardia* × *grandiflora*

*Gladiolus* × *hortulanus*

### Echinacea purpurea
Purple coneflower

Wherever you can grow black-eyed-susan (*Rudbeckia hirta*) you can grow purple cone-flower—which is to say, just about anywhere in North America. In fact, it resembles black-eyed-susan except for its purple color, more pronounced central cone, and slightly swept-back petals.

Native to midwestern prairies and eastern meadows, purple coneflower can withstand heat, humidity, and impoverished soil. It likes full sun but will tolerate light shade, which in fact protects the flower color from bleaching. Use it in mixed perennial borders and in mass plantings, especially in wildflower meadows. The 3-foot-high plants bloom in midsummer and have long stems suitable for cutting.

Purchase plants in containers from the nursery or divide well-established clumps to make new plants. Good varieties include 'Bright Star', an eye-catching rosy red, and 'White Lustre', with brilliant white flowers.

### Gaillardia × grandiflora
Indian-blanket

Wild species of gaillardias grow in impoverished sandy soil throughout North America. They relish open, sunny locations, tolerate high heat and humidity, and thrive in all but the heaviest clay or water-logged soil—making them fine candidates for an easy garden. Their fringed, 3- to 4-inch daisy flowers are usually red with yellow tips surrounding a prominent dark brown eye. Plants grow 1 to 3 feet tall depending on variety. They flower at midsummer, but removing faded flowers will prolong the display into fall.

Use gaillardias as highlights in mixed beds and borders. Dwarf varieties such as 'Goblin' can be used for edging and as drifts in rock gardens. Tall types, such as 'Burgundy', make dramatic sweeps in wildflower meadows.

A tall related species, *G. pulchella* (3 feet) is an attractive annual that flowers within 75 days after seed is sown directly into the garden. Look out for Monarch, an award-winning single-flowered color mixture, and 'Lollipops',

a series of double-flowered annuals that come in yellow, orange, and maroon. Both these and the tall perennial kinds are excellent for cutting. Buy plants in containers.

### Gladiolus × hortulanus
Gladiolus

Even if you live where the ground is frozen in winter—and the tender corms of gladiolus with it—you can still enjoy these flowers without digging up the corms each autumn. The bulblike corms (actually thickened, modified stems) are so inexpensive that it's no hardship to buy fresh ones each spring to replace the winter casualties. Just be sure to wait until severe frosts are over before planting them.

Native to semidesert areas of South Africa, gladiolus is no stranger to summer heat. It also tolerates a wide range of soils, provided that these have full sun and good drainage.

Applying high-phosphorus fertilizer in spring and watering during dry spells will produce the tallest, strongest, flower spikes. These spikes, studded with ruffled, 3- to 4-inch flowers, form a towering flower cluster that can measure 3 feet long on a 5-foot-high plant. Tall and top-heavy, the flowering stems may need staking. The bold, bright color range includes red, yellow, orange, peach, pink, purple, as well as white and bicolors. There is even a green gladiolus, popular with flower arrangers. Leaves are spiky like iris leaves.

For continuous color, stagger the spring planting of corms at two-week intervals. Cover corms with at least 4 inches of soil (deeper planting sometimes helps corms survive freezing winters). Gladiolus starts to bloom in midsummer and continues until fall frost. Use it in mixed beds and borders wherever a strong vertical accent is desired. It makes an especially good companion to tuberous dahlias. To have enough flowers for cutting, consider planting a row of gladiolus in a separate cutting garden or between rows of vegetables.

*Helianthus × multiflorus*
'Flore Pleno'

*Heliopsis helianthoides scabra*

*Hemerocallis* hybrid

Gladiolus specialists have produced many spectacular hybrids, which you can order through their mail-order catalogs. Alternatively, stick with old standbys such as 'Her Majesty' (violet blue), 'Peter Pears' (glowing orange), and 'Intrepid' (dramatic scarlet). These are available as corms from local garden centers.

## *Helianthus × multiflorus*
Perennial sunflower

Perennial sunflower is a hybrid of two prolific native North American sunflowers—annual sunflower (*H. annuus*) and thinleaf sunflower (*H. decapetalus*). An aggressive and abundantly flowering hardy plant, it sends up a towering pillar of golden yellow, 4-inch sunflowers in midsummer. Especially abundant is the full-double 'Flore Pleno'.

Plants thrive in full sun in any well-drained garden soil. They have such a commanding presence in mixed borders

that one plant is usually sufficient to draw attention. Use it at the back of the border or next to a picket fence, where the arching flower stems can cascade over and through the slats. It also looks fine in arrangements. Buy container plants from local nurseries, or divide established clumps in spring or fall.

A related species, swamp sunflower (*H. angustifolius*), also makes an attractive tall highlight for perennial borders. Its golden yellow, 4-inch flowers bloom in fall. Swamp sunflower will naturalize successfully in wet meadows or low-lying soils.

## *Heliopsis helianthoides scabra*
False-sunflower

In any search for easy-to-grow flowering plants, it is a good policy to first consider those that originated as North American wildflowers, and that are widely distributed. An eye-catching native, false-sunflower is not only hardy into Canada, but does equally well in southern gardens. The 4- to 5-foot bushy plants have

dark green, serrated leaves and produce a profusion of 3-inch golden yellow daisylike flowers in late summer.

Plant in full sun or partial shade in any reasonably fertile, humus-rich soil, and keep watered during dry spells. Use in mixed borders, especially as a background highlight; excellent for cutting.

Seed germinates reliably when sown directly into the garden, although the plants will not bloom until the second season. Local nurseries usually sell false-sunflower in containers, especially the varieties 'Incomparabilis', a semidouble with dark flower centers, and 'Summer Sun', a full double. Plants are easily divided in spring or fall.

## *Hemerocallis* hybrids
Daylily

Few garden perennials are easier to grow, have a longer blooming season, or are more decorative in the landscape

than hybrid daylilies—perhaps the ideal plants for carefree gardens. They withstand heat and drought as easily as they shrug off freezing cold and humidity. Although they prefer moist and fertile acid loam soil, they grow contentedly in dry, alkaline soil as well. They are true survivors—virtually foolproof and so easy to transplant that you can move them even when in full bloom. About the only condition they dislike is poor drainage.

Their botanical name, *hemerocallis*, means "beauty for a day," referring to the fact that each lilylike flower lasts less than 24 hours. However, nature compensates by crowding each flowering stem with dozens of buds. A modern hybrid may produce as many as thirty blooms on a single stem, with a total of up to nine hundred flowers on a well-established plant.

Plants grow to 3 feet high, creating an attractive cluster of arching, sword-shaped leaves. Flower colors are extensive, including shades of yellow, orange, red, apricot, peach, mahogany, lavender, and even green, plus bicolors. About the only colors missing are pure white and sky blue.

*Hibiscus moscheutos*
'Southern Belle'

*Hosta*

*Iris ensata*

Daylilies brighten beds and borders and succeed well in mass plantings on steep slopes. Their vigorous underground rhizomes and dense weave of fibrous roots hold soil in place, crowding out even the most persistent weeds.

Buy plants in containers from local nurseries, or increase established plants by root division. Any sausage-shaped root section can be transplanted to create a new plant. Thick clumps that bloom sparsely as a result of overcrowding can be rejuvenated by digging up plants, dividing their roots, and replanting them approximately 1 foot apart.

Two excellent varieties are 'Hypericon', an exquisite large-flowering yellow that grows 3 feet high, and 'Stella de Oro', a foot-high dwarf with orange flowers. Either is a perfect choice for modern gardens, providing color continuity from year to year with a minimum of maintenance.

## Hibiscus moscheutos 'Southern Belle'
Hardy hibiscus

The size of hardy hibiscus flowers is astounding—up to 12 inches across, or twice the size of tropical members of the genus. The largest-flowered of all garden perennials, *Hibiscus* species grow wild in swampy soil along the eastern seaboard. They have been hybridized almost beyond recognition, now flowering continuously from midsummer to fall frost and producing much larger flowers in a wider range of colors.

Hardy hibiscus reaches 6 feet in height, growing bushy if the main stem is snipped off to encourage side-branching. It prefers moist, humus-rich loam soil in full sun and will withstand high heat and humidity if watered weekly during dry spells. Use hardy hibiscus as the centerpiece of an island bed, as a background in a mixed flower border, or to enhance the margin of a pond or stream.

The award-winning 'Southern Belle' hybrid produces disk-shaped flowers in white, pink, and crimson, most with a contrasting darker eye and a powdery yellow mass of stamens extending prominently from the center. Like other cultivars of hardy hibiscus, 'Southern Belle' is easily raised from seed. Soak the bullet-hard seeds overnight to soften their coats and speed germination; start plants indoors and set them out eight weeks later, after frost danger has passed. Or buy plants in containers at the nursery.

## Hosta hybrids
Hosta, plantain lily

The *Hosta* genus, native to Japan, includes many species and varieties noted for decorative leaves that form a weed-smothering ground cover, especially in light shade. A bonus is the slender flower stems that appear in summer, topped with pendant bell flowers, usually white, purple, or lavender. Leaf shapes vary; some are slender and lancelike, others broad and paddle-shaped. Leaf texture can be smooth or exotically ribbed and puckered, accentuating the leaf veins. Hostas are hardy into Canada; and they survive heat well into the southern United States, provided that the soil has high humus content and remains moist.

*H. sieboldiana* (3 feet high) has enormous heart-shaped blue leaves more than 2 feet long and prominent white flower spikes that extend another foot above the leaves. The variety 'Frances Williams' was rated the best hosta in a recent poll of American Hosta Society members. It has heavily textured blue-green leaves and creamy yellow leaf margins.

Use hostas as an edging for woodland paths or wherever a weed-free area is needed under trees. The large-leaved kinds can be used as highlights in mixed perennial borders. Buy them in containers at garden centers.

## Iris ensata; I. kaempferi
Japanese iris

Unlike its cousins the bearded iris (*Iris* × *germanica*) and Siberian iris (*I. sibirica*), Japanese iris tolerates boggy soil so successfully that it can

*Iris × germanica*

*Iris sibirica*

be grown with its roots permanently immersed in shallow water. It also has a much flatter flower head than the other two. Given moist, humus-rich acid soil (it need not grow in water), this hardy perennial can thrive virtually anywhere in the United States, including the Deep South.

Japanese iris flowers in late spring or early summer. Its color range is mostly shades of blue and purple, with some white and two-tone varieties. It is popular for Japanese gardens, planted alone in clumps or at the edge of a pool or stream. It is also useful as a highlight in mixed perennial borders, preferably in clumps of a single color. Plants grow to 3 feet high and spread by rhizomes, swollen underground stems. Buy dormant rhizomes from mail-order catalogs, or purchase plants in containers from local nurseries.

Japanese iris is a perfect companion to yellow flag iris (*I. pseudacorus*), which also grows in moist soil and tolerates boggy sites.

## Iris × germanica
Bearded iris

A good indicator of whether bearded iris will grow easily in your area is whether corn will grow there. Like corn, this iris hybrid thrives in fertile, well-drained loam soil in full sun. No other perennial has such an extensive color range: red, white, blue, yellow, orange, peach, apricot, purple—even black and green—along with hundreds of two-color combinations.

Its decorative, bright green, sword-shaped leaves grow 6 inches to 2½ feet high, depending on the cultivar. Strong, erect flower stems tower another foot or more above the leaves. The large ruffled flowers, which open early in the summer, have an intricate petal arrangement: One set of large tongue-shaped petals arches down, while another set of petals arches upward around a conspicuous set of yellow stamens called a beard. Flowers have a spicy fragrance.

Use bearded iris in beds and borders, either in bold groups of one color or in a rainbow mixture. Dwarf varieties, such as 'Blue Denim', are good in rock gardens and along borders. Bearded iris forms rhizomes, similar to bulbs, which grow into replicas of the original plant. Clumps of plants form quickly and may need dividing after three years. Seeds are less likely to produce true replicas and, anyway, take too long to produce plants of flowering size. Purchase dormant rhizomes from mail-order sources, or buy plants in containers at nurseries. In late summer after blooming, you can cut rhizomes from established plants to increase your collection.

Top-rated bearded iris varieties include 'Tollgate', a blue and white bicolor; 'Beverly Sills', a bright, glowing pink; and 'Rich Reward', a large yellow. These are especially good for cutting. Although bearded iris tolerates heat and drought, it is best watered weekly during dry spells. It also appreciates a high-phosphorus fertilizer in fall after the leaves die down and in spring before flowering.

## Iris sibirica
Siberian iris

If any flower belongs in a list of top-ten easiest-to-grow perennials, it is Siberian iris. The hardiest and easiest-to-grow of all iris species, it is also heat tolerant and thus a good iris even for southern gardens. The white, blue, or purple flowers, which appear in late spring, are smaller but more refined and free-flowering than those of bearded iris, the usual favorite. The leaves, too, are more ornamental—slender and green in summer, turning a lovely golden and bronze hue in autumn.

Plants grow to 3 feet high in full sun, forming dense clumps of slender foliage. Although they tolerate poor, dry soil, they flower more freely under moist, humus-rich, slightly acidic conditions.

Use Siberian iris in clumps with other perennials in beds

*Lavandula angustifolia*

*Lilium* hybrid 'Enchantment'

*Lunaria annua*

and borders. The blues make good companions for pink and red Oriental poppies and purple foxglove. Mass plantings look magnificent along a pond margin or stream bank. They are among the finest perennials for cutting.

Purchase dormant rhizomes from mail-order specialists for spring or fall planting, or buy container plants from a local nursery. Established clumps are easily divided after flowering in summer or fall. The varieties 'Tealwood' (deep violet) and 'Sea Shadows' (turquoise-blue) are especially beautiful.

## *Lavandula angustifolia*
### English lavender

If you asked a group of gardeners what flower they would grow if they could choose only one, English lavender would likely be the overwhelming favorite. It not only blooms continuously from early summer to fall frost,

thriving in a wide range of soils in sun or partial shade, but it produces one of nature's most pleasant fragrances. Its leaves and flowers, steeped in boiling water, also make a refreshing herbal tea.

Plants have narrow leaves that grow in mounded, 3-foot-high clumps. Violet blue, mauve, or white flower spikes rise high above the foliage. Use English lavender in mixed beds and borders or as a low hedge to line a path or driveway. The long, strong stems are excellent for cutting.

Grow plants from seed started indoors, or take cuttings from established plants and root them in moist potting soil. Container plants are plentiful at nurseries. After the first frost in autumn, cut the stems to within a few inches of the soil line and cover the plants with a layer of organic mulch to help them survive winter freezing.

The 'Hidcote' variety has the deepest violet blue coloring. French lavender (*L. stoechas*) has purple, shorter flowers and is not so showy, but is favored for coastal gardens and for hot, dry areas with mild winters, such as central California.

## *Lilium* hybrids
### Midcentury hybrid lilies

It might once have been stretching the term *easiest* to include a garden lily in a listing of this kind, since lilies were notoriously temperamental. However, modern hybridizing has produced some dependable varieties that even the authoritative *Horticulture* magazine has pronounced "as easy to grow as daffodils."

The objects of this enthusiasm include some of the Asiatic hybrids, particularly the 'Midcentury' color mixture. These hardy plants will thrive in sun or shade in fertile, humus-rich soil that should be watered weekly in the absence of rainfall.

The 5-inch wide, star flowers of 'Midcentury' hybrids all face up rather than down as most lilies tend to, on sturdy 2- to 3-foot stems. The color range includes red, orange, yellow, pink, and white. These lilies are impressive highlights in mixed perennial borders,

and their long stems make them good for cutting. They will become naturalized in woodlands and meadows, where their only serious pests are deer, who eat the flower buds. Purchase bulbs by mail or at local garden centers; plant in fall or spring.

## *Lunaria annua*
### Moneyplant

An old-fashioned hardy biennial from southern Europe, moneyplant has silvery, papery seedpods that have long been appreciated by flower arrangers. Curiously, its lovely flowers have stayed out of the limelight. These begin their display in spring as 3-foot-tall purple clusters resembling old-fashioned phlox. There is also a white-flowered cultivar, and another purple one with variegated, deeply serrated, heart-shaped leaves that is destined to become popular once word spreads.

Plants are not fussy about soil and will grow even in stony ground or between cracks in flagstone. Give them full sun or partial shade, preferably in moist soil. One easy

*Lycoris squamigera*

*Monarda didyma* 'Cambridge Scarlet'

way to establish a clump is to sow seeds directly into the garden, where they can be left to mature and, eventually, to self-sow.

Moneyplant is a favorite for herb and cutting gardens. Use it sparingly in mixed borders and generously in naturalized settings such as woodlands or wildflower meadows. This plant does not bloom until the second season, when flowers are followed by the seedpods for which it is named. For flowers in the first season, buy plants in containers at a nursery.

## Lycoris squamigera
Naked-ladies

There are two kinds of flowering bulbs popularly called naked-ladies—*Amaryllis belladonna,* from South Africa, and *Lycoris squamigera,* from South America. In appearance they are indistinguishable, although *Amaryllis belladonna* blooms mostly in fall and is killed by frost, while *Lycoris squamigera* blooms in summer and is winter hardy.

Indeed, it is difficult to believe that tropical-looking plants such as *L. squamigera* can survive a single frost. Yet they not only survive in cold-winter regions but thrive there. These perennial flowering bulbs produce foot-wide clusters of pink or peach trumpet blooms, each bloom up to 6 inches long and flaring to 4 inches across. After bulbs take hold in spring, they sprout a cluster of dark green, arching, strap-shaped leaves that die down by midsummer; they are followed by 2- to 3-foot "naked" (leafless) stems from which the flowers emerge. After flowering, the bulbs go dormant to survive the winter.

Plants grow in any well-drained sandy or loam soil, in full sun. Use clumps of them as highlights in mixed beds and borders, massed in drifts, or naturalized in rock gardens and meadows. They are also suitable as container plants for decks, patios, or terraces.

Purchase bulbs in spring through mail-order bulb specialists; they are hard to find at garden centers.

## Monarda didyma
Beebalm

As with so many native species from the eastern United States, only outright abuse will stop beebalm from flowering spectacularly each year at midsummer. Tolerant of high heat, although not drought resistant, this vigorous plant likes moist, fertile soil and weekly watering during dry spells. Plant it in full sun or light shade.

Its tubular flowers, arranged in a crown 3 inches across, come in red, pink, purple, white, and orange. True to its name, it is highly attractive to bees as well as to butterflies and hummingbirds. Beebalm is a member of the mint family; its pleasantly fragrant, gray-green serrated leaves lend the pungent, smoky flavor to Earl Grey tea.

Plants grow to 3 feet high, forming a mounded clump with the flowers held high above the foliage. Use beebalm as a bold highlight in mixed perennial borders, dividing plants every three years to

keep them from becoming invasive. Popular for adding dramatic color to herb gardens, beebalm will also naturalize along streams and beside ponds. Its long stems make it good for cutting.

The most widely planted varieties are 'Cambridge Scarlet', an intense red; and 'Croftway Pink', a rosy pink. Purchase plants in containers locally. Alternatively, start your own from 4-inch tip cuttings taken before flowering, from root division after flowering, or from seeds sown in spring or summer.

## Myosotis scorpioides
Forget-me-not

Many plants with small blue flowers are referred to as 'forget-me-not', but *M. scorpioides* is the true one. Its masses of tiny, cheerful sapphire blue or pink flowers—which appear each spring on bright green, spear-shaped leaves—form compact plants. Although forget-me-not likes cool weather, there is hardly anywhere in North America it

*Myosotis scorpioides*

*Narcissus* 'Roulette'

*Oenothera pilosella*

will not flower, provided that plantings are timed to produce flowers when nights are still cool. Forget-me-not will grow in sun or light shade, preferring cool, moist, humus-rich loam soil.

Use forget-me-nots to fill beds and borders or as an edging—particularly around ponds, pools, and stream banks. An especially striking way to plant them is between tulips, which stand up tall amid their mist of small flowers. Another stunning location is at the base of a wisteria vine, where the lavender blue of the wisterias and the deeper blue of the forget-me-nots complement each other.

The genera *Brunnera* and *Anchusa* (perennials) and *Cynoglossum* (annual) contain plants sometimes called forget-me-nots that are just as easy to grow. The summer-flowering *Cynoglossum amabile* (Chinese forget-me-not) tolerates heat better than the others.

## *Narcissus* species and hybrids
Daffodils

Synonymous with dependability, daffodils bloom faithfully early each spring, year after year, and if conditions are right, become naturalized in the garden. Even rodents and deer will not touch them, since they are distasteful and indeed poisonous. The bulbs send up clumps of narrow, arching leaves and stout flowering stems topped with trumpet-shaped flowers in mostly yellow and white. They will multiply rapidly if planted in humus-rich soil and fed with a high-phosphorus fertilizer; apply the fertilizer in spring before the flowers bloom and again after the first frost in fall.

The most popular hybrid varieties include 'King Alfred' (golden yellow), 'Ice Follies' (white with a lemon yellow frilly trumpet), and 'Red Marley' (deep yellow with an orange-red cup). All grow to 2 feet high. Plant daffodils in light shade or full sun in beds and borders, in drifts under deciduous trees, at the edge of a lawn, or on slopes. They are excellent for cutting and can be grown in containers.

Species daffodils (wild kinds) are especially good for naturalizing. Some good choices are *N. minimus,* a miniature daffodil (6 inches); *N. jonquilla,* a highly fragrant daffodil with small flowers in clusters and reedy leaves (2 feet); and pheasant's-eye daffodil (*N. poeticus,* 2 feet), whose grassy leaves and abundance of slender flower stems are crowned with 3-inch fragrant white flowers with red-margined cups.

Buy bulbs by mail or from local garden centers any time after September 1, for fall planting. Cover with 6 inches of soil. Daffodils generally need at least 10 weeks of cold weather to ensure that they will come up in spring. After flowering, let the leaves stay on for at least eight more weeks to help plants make new bulbs. Cut the leaves only after they turn brown and wither away.

## *Oenothera pilosella*
Sundrops, eveningprimrose

Few yellow-flowering plants are as easy to grow in a wide range of soils—including impoverished soil—as the various species of evening-primrose native to North America. In the wild these grow contentedly on sand dunes and slag heaps, often becoming the first plants to take hold in soil ravaged by fire or strip mining.

Although several species of eveningprimrose are grown in home gardens, including *O. missourensis* (a trailing, untidy plant) and *O. speciosa* (a low-growing pink species that becomes naturalized at the blink of an eye), the most popular is *O. pilosella,* commonly called sundrops. Its shimmering, cupped 2½-inch yellow flowers cover the two-foot spreading plants so profusely that they completely hide the foliage. Plant *O. pilosella* in moist or dry soil in full sun. Use it to edge beds and borders for a two- to three-week splash of color in midsummer. Buy plants in containers at a nursery or garden center and propagate more by dividing mature clumps from the garden.

*Paeonia lactiflora*

*Papaver orientale*

*Phlox subulata*

### Paeonia lactiflora
Herbaceous peony

Although neither as carefree nor as widely adapted as some other flamboyant perennials, peonies are nevertheless surprisingly reliable given the size and quality of their exotic, round flowers that look like huge scoops of ice cream. In a sunny spot and fertile, loam soil that is watered during dry spells, peonies form bushy, leafy clumps up to 4 feet high. Topping them are gorgeous, fragrant pom-poms 8 inches across in white, pink, or red.

Peonies do best where winters are cold. They will tolerate light shade, but are heavy feeders that benefit from deeply dug, humus-rich soil and high-phosphorus fertilizer. They flower mostly in early summer.

Use peonies as highlights in mixed flower borders, or plant them as a hedge to line a driveway or path. Their long stems make them excellent for cutting. Peonies can be purchased in bare-root form by mail, or in containers from local nurseries. Established clumps can be divided and replanted with the tops of the roots set just 1 inch below the soil surface. Any root section with three to five "eyes" (growing points) is the right size for transplanting.

A strain of large-flowered hybrids called 'Estate Peonies', developed by a peony breeder near Chicago, is especially beautiful. 'Pink Parfait' (gorgeous deep pink), 'Jay Cee' (double, bright red), and 'Raspberry Sundae' (snow white flecked with red) make a dynamic trio.

Tree peonies (*P. suffruticosa*) will grow to 10 feet high. They produce even larger flowers than herbaceous peonies, and may carry up to a hundred blooms per plant. Moreover, they are not as dependent on cold winters as herbaceous peonies. Give them cool, moist soil with "their feet in the shade and their heads in the sun," as the adage goes, and a sheltered location. Mulch after the ground freezes in winter.

### Papaver orientale
Oriental poppy

A hardier, more reliable perennial than Oriental poppy is hard to find. Its roots remain viable through the coldest winters, breaking dormancy at the first warming to flower spectacularly by the end of spring. Its huge, frilly flower heads—up to 10 inches across—shimmer on sunny days, making an outstanding display when planted in a mass of a single color.

Crimson with black petal markings is the most popular color, but the range includes white, orange, pink, purple, and two-tone combinations. A mound of powdery black stamens creates a bold contrast in the center. 'Beauty of Livermore' (4 feet tall) is a true Goliath, bearing crimson-and-black, bowl-shaped flowers as big as dinner plates. This and other tall varieties may need staking; dwarf 'Allegro' (2 feet) stays neat and compact.

Before flowering, plants form a cushion of attractive ferny foliage that turns brown and shrivels once flowering is over. If a companion such as black-eyed-susan is planted nearby, its billowing form will hide the browning foliage.

Oriental poppies prefer sunny loam soil with high humus content and good drainage. They grow like weeds from seed but will not bloom until the second year. For same-season flowering, buy plants in containers, but take care not to disturb the roots while transplanting or the plants will wilt and die. Established clumps may be easily divided.

Oriental poppies are short-lived in warm-winter areas such as Florida and southern California; Iceland poppies (*P. nudicaule*), grown as annuals, are more popular in these climates since they flower faster.

### Phlox subulata
Moss phlox

If you would like a flowering lawn, moss phlox will provide one more successfully than any other low-growing flowering perennial. In spring, this North American wildflower produces a solid sheet of flowers so dense that they completely hide the gray-green,

*Physostegia virginiana*

*Scabiosa caucasica*

*Sedum spectabile* 'Autumn Joy'

needlelike evergreen leaves. The small, starry, ½-inch flowers are white, pink, red, or blue. The plants spread out instead of up, creating a 4-inch-high carpet ideal for covering slopes, edging beds and borders, cascading over dry walls, or creating drifts of color in a rock garden. All they demand is sunny, well-drained soil.

Many varieties are available, including 'Crimson Beauty', with bright red flowers, and 'Millstream Jupiter', a dark blue. A related species, blue phlox (*P. divaricata*), is a wonderful flowering ground cover for light shade. In fertile, moist soil, it forms solid sheets of blue.

Purchase plants in containers from local nurseries and plant in early spring. Established clumps are easily divided to create more plants.

### Physostegia virginiana
Obedientplant

Valued for its late-season flowering (end of summer, early autumn), the obedient-plant gets its name from the way its flowers can be pushed to one side or the other and

stay tamely in place. Plants grow 2 to 3 feet high, producing multitudes of stiff flower stalks and spear-shaped leaves, topped by clusters of tubular flowers that resemble snapdragons. These are pink or sometimes white.

The plants are hardy into Canada and tolerate heat, drought, and air pollution. In moist soil and full sun they can become aggressive, spreading rapidly and crowding out adjacent plants. Use them sparingly as a late-flowering highlight in mixed beds and borders, or grow them for cutting.

The cultivar 'Bouquet Rose' has beautiful deep pink flowers. 'Variegata' has willowy, cream-and-green leaves.

### Scabiosa caucasica
Pincushionflower

Easy-to-grow blue flowers of any decent size are hard to find; pincushionflower is one of the best. The ruffled, flat flower heads are powder blue

and up to 4 inches across, with a conspicuous white crest (the pincushion) at the center. The slender, wiry flower stalks grow 3 feet high from a clump of narrow green leaves.

The hybrid 'Clive Greaves' has especially large flowers. A white form, 'Miss Willmott', is also available. These members of the perennial *S. caucasica* species are much more beautiful than the related *S. atropurpurea*, an annual whose flowers are only half the size.

Buy plants in containers from a local nursery, or divide established clumps in spring. Pincushionflower likes full sun and well-drained loam or sandy soil. Group several plants together (in a mixed perennial border, for instance), since one plant doesn't make a sufficiently bold statement. You'll also want to have enough for cutting.

### Sedum spectabile
Stonecrop

Always high on any list of worthy perennials, stonecrop is not only virtually indestructible but produces an

extravagant flower show in late summer and early autumn, after many other perennials have finished for the year.

Two varieties are exceedingly beautiful: the bright pink 'Brilliant' makes a cluster of succulent 2-foot stems topped by flat flower clusters, up to 6 inches across, that are irresistible to butterflies; 'Autumn Joy', a hybrid almost identical to 'Brilliant', has rosy red flowers that gradually change to bronze as the flowers dry. The flower heads persist until late December and are attractive in dried arrangements. Use stonecrop in beds and borders, in container plantings, or as a companion to ornamental grasses such as fountaingrass.

Stonecrop grows throughout the United States and Canada. It stands freezing winters, tolerates poor soil, and is not particular about moisture. It thrives in sun and light shade. It is readily available in containers, and once established can be further increased by dividing the clumps.

*Stachys byzantina*

*Tulipa kaufmanniana* 'Overture'

*Yucca filamentosa*

### *Stachys byzantina*
Lamb's-ears

Although better known for their silver, velvety lancelike leaves, lamb's-ears also have lovely erect flower spikes studded with white or purple blossoms. These ground-hugging plants are extremely hardy, almost evergreen, maintaining an attractive appearance until harsh freezing. If watered weekly during dry spells, they will tolerate high heat and humidity as well. They prefer full sun and fertile, humus-rich soil, but with frequent watering will also accept poor, sandy soil.

Use lamb's-ears as highlights in mixed flower borders, as an edging for paths, or wherever you need a ground cover. In flower at midsummer the plants stand 3 feet high, but before and after flowering they form rosettes just 3 to 6 inches high. 'Silver Carpet', a sterile cultivar, stays low because it doesn't flower. Lamb's-ears are sometimes used in herbal wreaths; the fresh or dried flower stems add a silvery note. Divide plants every three years to keep them from invading nearby perennials.

### *Tulipa* species and hybrids
Tulips

Most gardeners choose hybrid tulips for their drama, but it is the more unassuming species tulips—tulips that breed true from seed—that will lead the most carefree life in your garden. Unlike hybrids, which become less vigorous after the first year, species tulips perform consistently year after year, and if you let them they will become permanent residents. Plant bulbs in fall, just before the first frost, and they will come up after daffodils in spring, beating hybrid tulips into flower by several weeks.

Most species tulips (also known as botanical tulips) are low growing—6 to 12 inches high depending on variety. One particularly handsome species is *T. kaufmanniana*, called waterlily tulip because its yellow, white, or red flowers open flat like waterlilies. Another is *T. fosterana*, especially the 'Red Emperor' variety, which displays huge, shimmering red 6-inch flowers with black throats.

Crossbreeding has created many highly developed hybrids, which are classified in several categories. These include the lily-flowered types (with elegant pointed petals), parrots (with large, bizarre, feathered petals), and fringed types (with frilly petals). Other types include the popular 'Darwin' hybrids, which have extra large, urn-shaped flowers in a stunning color range—shades of yellow, orange, pink, red, and purple along with white and bicolors. Longer lasting by as much as five days are 'Cottage', 'Triumph', and regular 'Darwin' tulips (including the famous egg-shaped black 'Queen of the Night').

Tulips look their best in beds and borders; the more compact species tulips are also attractive in container plantings. For the best selection, buy bulbs as soon as possible after September 1 and plant them no later than Christmas, covering them with 6 inches of fertile, humus-rich sandy or loam soil in full sun. They will also tolerate light shade, where indeed the flowers will last longer away from the bleaching effects of sunlight. Protect bulbs from rodents by scattering mothballs or rodent-repellent flakes in the bed, or by covering the planting area with chicken wire.

Tulips flower most reliably and profusely in areas with cold winters. In warmer climates, force them into bloom by giving the bulbs 10 weeks of cold storage in a refrigerator before planting them outdoors. To encourage reflowering the next season, pinch off faded flowers and allow the leaves to die down naturally. Unless faded flowers are removed, seedpods may form, robbing bulbs of energy. Adding high-phosphorus fertilizer in early spring and late summer will also strengthen plants.

### *Yucca filamentosa*
Desert-candle

Native to the Arizona desert, desert-candle thrives in hot, dry conditions, yet is also a surprisingly cold-tolerant plant—a rare combination for a perennial. Evergreen and deceptively tropical in appearance, desert-candle has sword-shaped leaves that grow 2 to

*Albizia julibrissin*

*Amelanchier arborea*

3 feet long in a spiky rosette. In early summer a tall, slender flower spike—resembling a giant asparagus spear—towers some 6 feet above the leaves and erupts in a shower of pure white, nodding flowers.

Desert-candle prefers full sun in moist or dry soil. It can survive long periods of drought. Plant it in small groups as part of a mixed perennial border or in mass in a rock garden, along a dry slope, or in a desert landscape. Avoid planting it close to paths or where children play, since the leaf tips are sharp.

Mature plants produce offsets called pups, which you can easily remove to establish new colonies. Otherwise, buy plants in containers.

## FLOWERING SHRUBS, TREES, AND WOODY VINES

Woody plants add substance and height to the garden. When not in bloom, they provide a green backdrop for other flowering plants. Most grow slowly, so start them from ready-grown stock.

### *Albizia julibrissin*
Silk tree

Few deciduous flowering trees can survive the hot, dry summers of the Southwest, the freezing winters of the Northeast, and the damp summer fog of the West Coast as successfully as the silk tree. Introduced from China, it should not be confused with the less-desirable mulberry tree, brought from Asia to feed silkworms.

Young trees purchased in containers from local nurseries grow quickly into graceful, spreading canopies of bright green, delicate tropical-looking leaves. These let through enough filtered sunlight to permit grass to grow right up to the strong, slender trunks. Clusters of brilliant

pink flowers, resembling shaving brushes, last for several weeks in midsummer and are followed by brittle pods containing bean-shaped seeds. These germinate easily, often sprouting into seedlings that grow in the shade of established trees.

Silk tree grows in a wide range of soils, provided that drainage is good. Plant in full sun and water during dry spells. Use it as a lawn highlight, beside a patio, or as a street tree.

### *Amelanchier arborea*
Sarvis tree

Not many native North American trees ever receive an Award of Merit from the prestigious Royal Horticultural Society of Great Britain. Such an honor, coupled with the distinction of being easy to grow, is exceedingly rare. A lovely ornamental that grows wild from Maine to Florida, sarvis tree heralds spring with a blizzard of snow white, crabapplelike blossoms that appear even before crabapple trees bloom. In summer it transforms itself into a cool

green arbor that yields masses of sweet, edible, blueberrylike fruit clusters. In autumn, its spear-shaped leaves turn golden, falling away by winter to reveal a delicate tracery of branches.

Sarvis tree, which grows to 40 feet high, can be started from bare-root, containerized, or balled-and-burlapped stock. It prefers moist, fertile, humus-rich acid soil in sun or partial shade. Show off this plant's magnificent flowering by using it as a lawn accent or as a contrasting element against fences or between evergreens, especially along a house foundation.

### *Campsis radicans*
Trumpet-creeper

If all the other plants in the world were to perish from pollution and drought, it is likely that this tenacious flowering vine would still survive, flowering its heart out. A hardy, indestructible North American native, it grows wild along the East Coast from Maine to Florida, producing attractive leaves similar to those of wisteria and 6-inch-wide clusters of orange-red,

*Campsis radicans*

*Caryopteris × clandonensis*

*Cercis canadensis*

trumpet flowers that attract hummingbirds more successfully than any other plant. Flowering continuously all summer until fall frost, plants may grow 10 feet in a season.

There is a hardy yellow-flowered variety, 'Flava', and a hardy larger-flowered, orange-red hybrid, 'Madame Galen', of which *C. radicans* is a parent. Plants grow easily from beanlike seeds and from cuttings, and are readily available in containers at garden centers. They thrive in sun or light shade.

### Caryopteris × clandonensis
Blue mist shrub

Blue is not a common color among deciduous flowering shrubs, particularly in late summer when blue mist shrub blooms. The lovely powder blue or deep blue flowers of this bushy shrub are pleasantly scented and clustered along slender stems like soft-bristled brushes. The shrub's upright stems have small, pointed leaves.

Blue mist shrubs usually stay below 5 feet in height, with an equal or slightly wider spread. Although they may be killed to the ground during severe winters, new stems will usually sprout from the roots in spring, to flower profusely on schedule. Purchase container-grown plants in bloom so you'll see exactly what shade of flower you are getting. Plant in sunny loam soil with good drainage, and water whenever a week goes by without rainfall.

The blue mist shrub works well in mixed shrub borders or in combination with perennials. It also makes an attractive hedge along driveways and paths. The flowering stems are delightful in arrangements. *C. incana,* a similar plant that is commonly available, is in fact a parent of this hybrid. However, it is not as hardy, large, or showy.

### Cercis canadensis
Eastern redbud

It really is extraordinary to see eastern redbud growing happily in the acid soils of the Northeast *and* in the alkaline soils of the Southwest. This small flowering tree usually produces masses of reddish pink, pea flowers on leafless stems, although white-flowered kinds are also available. The large, heart-shaped leaves that follow the flowers are dark green or purple depending on variety, changing to lime green or yellow in fall.

Native to open woods of the eastern United States, eastern redbud grows to 20 feet high. It prefers moist, well-drained soil in sun or light shade. Its western counterpart, western redbud (*C. occidentalis*), is almost identical in appearance; it thrives mainly in California. There is also a Chinese form (*C. chinensis*), a shrubby plant that grows only 10 feet high and has many stems instead of a main trunk.

Redbuds make good lawn highlights and are also popular for groves and woodland gardens. Transplant young

trees from containers or from balled-and-burlapped stock available at nurseries.

### Chaenomeles speciosa
Flowering quince

After a bleak winter, flowering quince is a welcome sight, lighting up the landscape like a beacon. Its cupped, rose-colored flowers are all the more vivid because they bloom on leafless branches.

An especially beautiful shrub for companion planting, flowering quince looks exquisite alongside white star-magnolia and yellow forsythia, which bloom at the same time in early spring. The color range includes pink, red, orange, and white. In autumn, yellow fruits the size of golf balls may appear. Although bitter if eaten raw, these make delicious jelly and jam.

Best bought in containers, plants grow to 10 feet high, although you can keep them below 6 feet by pruning the branch tips after the flowers fade. The natural shape is

*Chaenomeles speciosa*

*Clematis paniculata*

*Cornus florida* 'Rubra'

rounded and spreading, but shearing will shape plants into compact cushions or mounds.

Flowering quince tolerates poor soil in sun or partial shade, but flowers best in sunny, humus-rich loam. Plants that are overgrown can be rejuvenated by cutting the entire shrub to within 6 inches of the soil line, and then sprinkling the soil with a general-purpose fertilizer to encourage new growth.

Use flowering quince as you might azalea—as a lawn highlight, massed in the landscape, sandwiched between evergreens along foundations, or as a hedge. Branches in bloom are exquisite for cutting. Two especially attractive varieties are 'Rubra Grandiflora', with crimson flowers almost 2 inches across, and 'Toyo Nishiki', with apple-blossom-pink flowers.

## Clematis paniculata
Sweet autumn clematis

Many hybrid varieties of vining, or climbing, clematis are relatively easy to grow, provided that they have their

"feet in the shade and their heads in the sun," as the old gardening adage goes. Two easy-to-grow hybrids are the European-bred 'Jackmanii' (violet blue) and 'Nelly Moser' (pink and white), which both have star-shaped flowers up to 10 inches across that bloom in early summer.

However, it is the fragrant, white-flowered sweet autumn clematis, a North American native, that is the real survivor, tolerating poor, infertile, sandy soil that kills other types of clematis. In fertile, humus-rich soil it knows no bounds, covering fence posts, trellises, and chain-link fences with masses of fragrant, ½-inch starry flowers that bloom in early autumn and completely hide the foliage. Under the right conditions this plant will climb 10 feet in a year, rivaling trumpet-creeper (*Campsis radicans*) as a quick cover for arbors, pergolas, and gazebos. One mail-order company, in its enthusiasm for promoting this variety, called it "galaxy vine" and declared that it "grows so many flowers you could not possibly count them. . . a million or more on a single

vine. . . ." So abundantly does this vine flower that this is probably not the exaggeration it seems.

Plants flower the first year from seed, which is easy to grow. However, they are best purchased in bare-root form through mail-order specialists, since seeds are hard to find.

## Cornus florida
Flowering dogwood

On its native turf along the eastern seaboard, flowering dogwood is one of the most lovely and successful of hardy ornamental trees—yet outside North America it rarely does as well, due to its liking for sharp winters and warm, sunny summers.

The tiny and inconspicuous true flowers of this small deciduous tree are surrounded by decorative bracts (modified leaves) in white, pink, or rosy red, which hold their color for several weeks. The bracts precede the main foliage, carrying the ornamental display up to 30 feet skyward like a flock of butterflies. The oval, pointed

leaves have good fall color—usually reddish bronze—highlighted by red berries that are loved by songbirds. Dogwoods underplanted with azaleas, flowering together in spring, are among the most extravagant combinations in nature. And a single branch cut from a dogwood in bloom is a ready-made bouquet.

Flowering dogwoods prefer moist, humus-rich acid loam soil in full sun or partial shade. Buy small trees balled and burlapped and transplant them into well-drained loam soil in spring or fall. Young trees in the open should be protected from sun with tree wrap until well established, to prevent sunscald and damage by borers.

In recent years a virulent form of dogwood anthracnose fungal disease (also called lower-branch dieback) has been killing dogwoods; symptoms are spotty leaves and dying lower branches. Dogwoods under any environmental stress, including poor soil and dry conditions, appear to be especially susceptible. To keep a dogwood healthy and disease resistant, fertilize it in early spring with a balanced or high-nitrogen fertilizer, mulch

*Cotinus coggygria*

*Crataegus phaenopyrum*

*Forsythia × intermedia*

around the trunk with wood chips to conserve soil moisture, and water whenever a week goes by in summer without rainfall.

Several other dogwoods are widely planted, but none is a match for *C. florida* in beauty and resilience. The Korean dogwood (*C. kousa*) has the advantage of greater disease resistance, but it flowers a month later than the natives, thus missing the chance to participate in the floral fanfare of early spring. The western dogwood (*C. nuttallii*) has larger flowers, but doesn't provide quite the same spectacle and never survives long "back East," where hot, humid summers create too much stress. Similarly, the eastern dogwood flourishes in the West, but flowers less spectacularly than in its native habitat.

### Cotinus coggygria
Smokebush

Sure to evoke admiring comments from visitors when it reaches peak flowering in midsummer, *C. coggygria* becomes crowded with billowing flower clusters that resemble clouds of smoke. Actually,

these "flowers," which appear on the topmost branches, consist of masses of brittle stems that support inconspicuous true flowers. Once the true flowers have finished, the dried stems remain decorative for a month or more, lasting even longer when cut and used in dried-flower arrangements. Colors of the flowers and stems include pink, purple, and white.

Although this small, rounded, deciduous shrub will grow to 15 feet high, it can be kept low and bushy by heavy pruning. A well-pruned smokebush works well in mixed shrub borders, especially along a house foundation. It is also useful as a lawn highlight. The 'Purple Robe' variety displays rich purple foliage and purple-red flowers, contrasting well with other shrubs' green or silvery foliage.

Easily transplanted from either bare-root stock obtained by mail-order or from containers available at garden centers, smokebush prefers a sunny position and well-drained loam soil. The only serious drawback of this easy-to-grow shrub is its brittle branches, which are some-

times damaged by high winds. Do not confuse the smokebush, a Chinese native, with the American smoketree, *C. obovatus*. The latter is less desirable as a landscape plant, although its orange fall coloring is impressive.

### Crataegus phaenopyrum
Washington hawthorn

If you like ornamental cherries and crabapples, you will love the Washington hawthorn, which blooms in spring after cherries and crabapples have finished. Washington hawthorns are small flowering trees, usually long lived, that produce a myriad of small white, light pink, or rose pink blossoms on thorny branches. The dark green, serrated leaves contrast well with the ornamental red berries that ripen in autumn. Native to the eastern United States, it is one of more than one hundred hawthorn species found across the continent. It tolerates poor soil and polluted air. However, in areas planted with *Juniperus virginiana* (red cedar juniper), Washington hawthorns are susceptible to rust disease.

Transplant dormant young trees from bare-root, containerized, or balled-and-burlapped stock, preferably in early spring. Plants need full sun and good drainage. They grow up to 30 feet high, but can be kept low and compact by heavy pruning. Washington hawthorn can withstand crowding and will make a solid, impenetrable hedge if sheared regularly with hedge trimmers.

A related species, *C. viridis*, has pure white flowers. 'Winter King', a variety of *C. viridis*, produces one of the best berry displays that persist into winter. *C. laevigata* (English hawthorn) is less widely adapted and is best grown in the Northeast and Midwest.

### Forsythia × intermedia
Forsythia

Planted in sun or light shade, rich soil or poor, forsythia is so eager to bloom that even year-old rooted cuttings purchased from discount stores will flower the very next season. After five or six years when the plant becomes tall and untidy, you can take

*Hibiscus syriacus*

*Hydrangea paniculata* 'Grandiflora'

hedge trimmers (or even a chain saw), cut all branches to within a few inches of the ground, and watch it rejuvenate in a single season.

Forsythia is a popular shrub among home gardeners not only for its vigor but for its billowing masses of golden yellow flowers on leafless branches, which announce spring as exuberantly as crocuses or daffodils. Let forsythia grow unpruned for an informal look. Plant it either singly as a lawn highlight or in series as a hedge, with plants spaced 3 feet apart. The low-growing 'Arnold's Dwarf' cultivar works well as a ground cover for slopes. Good shrubs to plant with forsythia for simultaneous flowering are flowering quince (*Chaenomeles speciosa*) and star magnolia (*Magnolia stellata*).

Plants grow to 8 feet high and flower for up to three weeks in spring, provided they are not hit by a sudden severe freeze or warm spell. The small, oval, serrated green leaves sprout as soon as the flowers start to fade. Buy young plants in containers from local nurseries. For mass

plantings as hedges or ground covers, choose less-expensive bare-root plants from mail-order sources.

Forsythias are among the easiest shrubs to force into bloom for winter color. In early spring when the buds swell, cut 2- to 4-foot-long canes and place them in water indoors under bright light. Within a few days the buds should open in a cheerful prelude to spring.

### Hibiscus syriacus
Rose-of-Sharon

One of the most profusely flowering shrubs of late summer, rose-of-Sharon is also one of the last to leaf out in spring, prompting many a new gardener to wonder if winter has killed it. But if all else in the garden is thriving, this is most unlikely—in fact, this deciduous plant is usually the last to succumb to stress or neglect. Be patient, and your reward will be masses of 4- to 6-inch hibiscuslike flowers on upright, mound-shaped plants with maplelike foliage.

Flower colors include pink, white, and blue—some with contrasting red eyes. There

are both single and double forms. In recent years, plant breeders at the U.S. National Arboretum in Washington, D.C., have produced some exceptionally large-flowered hybrids, including 'Diana', a white-flowered variety, and 'Helene', which has white flowers with deep pink centers. However, old-fashioned 'Blue Bird' is still the favorite among gardeners.

Rose-of-Sharon tolerates a wide range of soils, including stony, sandy, and clay soils. However, for the most profuse flowers, set out plants in a moist, humus-rich soil. Plant in full sun or partial shade, and don't worry about hot weather—high temperatures don't faze rose-of-Sharon. Plants will grow to 12 feet, but look best if kept sheared with hedge trimmers to about 5 feet.

Buy plants in containers at the nursery and use them to line a driveway, hide a house foundation, or serve as an informal flowering hedge. Some nurseries offer plants with red, white, and blue strains grafted onto a single rootstock, for an all-American effect.

### Hydrangea paniculata 'Grandiflora'
Peegee hydrangea

Blooming in late summer, peegee hydrangea displays enormous white flower clusters—up to 12 inches long and almost as wide. Flowers remain decorative for weeks on this striking deciduous shrub, gradually changing to pink and finally to bronze as the petals dry to a parchment texture. The long, arching, woody stems can be cut for fresh arrangements when the flowers are in their prime, or for dried arrangements later on. The large, attractive leaves are spear-shaped, serrated, and heavily veined.

Use hydrangea as a focal point on a lawn, or as a backdrop for a mixed shrub border. A row of plants makes a magnificent informal hedge, which can be kept bushy by heavy pruning in autumn to within 6 inches of the soil line; this forces new growth up from the roots the following season. Unpruned plants will grow to 25 feet high, but normally become top-heavy and arch sideways at 10 to 15 feet. Whether you prune or

*Koelreuteria paniculata*

*Lagerstroemia indica*

*Lonicera sempervirens*

not, thin the many flowering stems to about six in total, so that energy is directed into creating extra-large flowers. Do this in early spring or autumn. If trained to a single stem, peegee hydrangea forms a charming small tree, with an umbrella canopy of leaves and flowers and a strong main trunk.

Plants thrive in sun or partial shade in a wide range of soils, provided that drainage is good. Humus in the form of peat, compost, or leaf mold will help keep the soil moist and will aid flowering.

### Koelreuteria paniculata
Golden-rain-tree

Performing well from Canada to Florida, the golden-rain-tree produces a fine, strong, straight trunk and a dense, spreading canopy of serrated leaves that help accentuate the bright yellow flower clusters. Even after flowering, the bold display continues with masses of papery seed cases resembling Chinese lanterns. These change from lime green to parchment brown as the season progresses.

Plants grow to 30 feet high and prefer full sun; they tolerate air pollution, heat, drought, and acid or alkaline soils. Transplant from balled-and-burlapped or container stock. Use as a lawn highlight or street tree; cut branches bearing dried seed cases for arrangements.

A related species, *K. bipinnata*, is similar in appearance but more colorful, with seed cases that turn bright pink. Although not hardy in the north, *K. bipinnata* is popular in Florida, the Gulf states, and southern California.

Don't confuse the summer-flowering golden-rain-tree with the golden-chain-tree, *Laburnum × waterii*, a spring-flowering tree that is less widely adapted.

### Lagerstroemia indica
Crapemyrtle

Without this spectacular deciduous flowering tree to provide vibrant color, many warm-climate gardens would look drab in summer. Hardy as far north as Philadelphia on the Atlantic Coast and into Canada on the Pacific Coast,

crapemyrtle is handsome in all seasons. Usually it has multiple stems with shiny, smooth gray bark, or sometimes fluted bark in older specimens. A canopy of rich green oval leaves emerges in late spring, followed by massive flower clusters resembling lilac blossoms, in red, pink, white, or lavender blue.

Plants prefer full sun and grow quickly, to 30 feet high, although some dwarf, bushy cultivars are also available. Crapemyrtles tolerate poor soil as long as it is well drained, but humus-rich soil produces the best flowering display. If plants are tip-pruned after the first flush of flowers, the branches will flower again. Fall leaf color can be dramatic shades of orange and red.

Many fine varieties have been developed by the U.S. National Arboretum, selected especially for their disease resistance, prolific flowering, and attractive bark. Some can be pruned with hedge shears to keep them compact and shrubby; these smaller cultivars are also suitable for growing in tubs.

Buy plants balled and burlapped or in containers from local nurseries. Use them

as sentinels at entrances, as lawn highlights, or planted in groves with a dark evergreen ground cover to help accentuate their polished bark.

### Lonicera sempervirens
Scarlet honeysuckle

Why anyone would want to grow the insipid yellow-flowered Japanese honeysuckle vine (*L. japonica* 'Halliana'), when this perfectly hardy North American native is more beautiful and drought tolerant, is a mystery. Scarlet honeysuckle grows wild throughout the northeastern United States from New England to Florida, flaunting masses of scarlet flower clusters composed of 3-inch trumpet blooms. Plants grow in full sun or light shade, scrambling upward some 10 feet a season even in poor, infertile soil.

Use scarlet honeysuckle as a vine to decorate arbors, trellises, or chain-link fences; or as a shrubby backdrop to a flower border. Buy bare-root plants from mail-order specialists, or in containers from local nurseries. Prune heavily in fall to keep them bushy and compact.

*Magnolia grandiflora*

*Magnolia × soulangiana*

*Malus floribunda* 'Snowdrift'

## Magnolia grandiflora

Southern magnolia

Native to North America, southern magnolia is a handsome evergreen tree with broad oval leaves and huge 10-inch pure white flowers that bloom in early summer. The waxy blossoms have an almost intoxicating fragrance and open flat like waterlilies, displaying a prominent, creamy white mass of powdery stamens at the center.

Southern magnolia is not reliably hardy north of Philadelphia on the Atlantic Coast, but on the Pacific Coast it thrives into Canada. Plants grow to 80 feet high in a fine pyramidal form. Even when not in flower, the large paddle-shaped, shiny, dark green leaves are decorative, often with brown, felty undersides. Plants prefer moist, humus-rich, acid soil in sun or shade.

Use southern magnolia as a lawn highlight or to line a driveway. In small gardens it can be trained flat against a wall. Buy plants balled and burlapped from nurseries.

## Magnolia × soulangiana

Saucer magnolia

A hybrid of species native to the Himalayas and mountains of western China, saucer magnolia thrives from central Florida and southern California well into Canada and is the best loved of all magnolias grown in North America. Its parent plants, *M. quinquepeta* (lily-flowered magnolia) and *M. heptapeta* (Yunnan magnolia), are each beautiful small trees that produce multitudes of gorgeous flowers in early spring. Yet this hybrid, developed by a French nurseryman in 1826, outdoes them both. It covers itself with unbelievable quantities of flowers that are usually deep pink on the outside and white on the inside, measuring up to 10 inches across when fully open.

The flowers open at the first sign of prolonged warming in early spring, before the leaves appear. If frost occurs while the flowers are open, they turn brown and die; but there are usually many years of good flowering between years of ruinous frost. The large, oval, pointed leaves are deep green in summer and turn brown in autumn.

Plants grow to 40 feet tall, usually with multiple stems. If a stem breaks off from ice damage or rot, the tree quickly sends up a replacement.

Transplant balled-and-burlapped trees into fertile, moist, acid soil that has been enriched with plenty of peat moss, leaf mold, or compost. Plants prefer full sun and tolerate air pollution. Use them singly as lawn highlights or plant them in groups of three. A single cut branch makes a bouquet. Many good varieties offer a color range from pure white ('Alba Superba') to rosy purple ('San Jose').

A related species, also worth garden space, is *M. stellata* (star magnolia). It grows like a shrub with masses of white flowers that are about half the size of other magnolia blossoms but bloom a week or two earlier.

'Majestic Beauty' flowers at an early age (three years, as opposed to six years for other types), has unusually large leaves, and produces many more flowers than the native species.

## Malus floribunda

Japanese crabapple

So many crabapples fall into the easy-to-grow category that it is hard to single out the best. Sargent's crabapple (*M. sargentii*), a white-flowered, small, spreading tree, is certainly one of the most eye-catching, with pure white blossoms so densely packed that they light up the landscape like a billowing cloud. However, the larger Japanese crabapple has a slight edge: Its blossoms are tinged pink, and it has a more rounded habit, making it a better choice.

Japanese crabapple is both vigorous and long lived; one specimen planted by French painter Claude Monet at the turn of the century still puts on a remarkable show in his famous gardens each spring. On the East Coast, Japanese crabapple grows well from Canada to the Florida border; on the West Coast and in other gardening areas, it thrives wherever apples can be grown. Its small, red, ornamental fruits appear in autumn. These are loved by songbirds and many are edible in jams.

*Polygonum aubertii*

*Prunus cerasifera* 'Atropurpurea'

*Pyrus calleryana*

Plants grow to 30 feet high, with a rounded crown. They tolerate a wide range of soils, although they prefer sunny locations that are slightly acid and rich in humus. Transplant them from containerized or balled-and-burlapped stock.

Many handsome hybrid crabapples are readily available from local nurseries and mail-order specialists, some with good resistance to the diseases that afflict them, such as fireblight. Two hybrids that have become exceedingly popular for their appearance and disease resistance are 'Dolgo', which bears masses of carmine red blossoms; and 'Red Jade', a weeping form with rosy red blooms that are followed in autumn by numerous deep red ornamental fruits. 'Red Jade' can be grown in containers.

## Polygonum aubertii
### Silverfleece vine

Were there to be a contest to determine which flowering plant grows the fastest, the silverfleece vine would probably win hands down. It can grow 15 feet or more in a single season and can cover a small barn within two. Its

hardy, twining stems grasp any kind of support, producing slender, oval leaves and myriad slightly fragrant small white flowers. These are borne in dense clusters that create the fleece effect for which the plant is named.

Plants grow in any type of soil (even pure sand) and can survive long periods of drought. Just ensure full sun and a strong support—such as a trellis or an arbor.

A pink-flowered species, *P. reynoutria,* is popular in frost-free areas such as southern California.

## Prunus cerasifera 'Atropurpurea'
### Purple-leaf plum, pissard plum

The large family of ornamental *Prunus*—which includes cherries, peaches, plums, and almonds—has some incredibly lovely springflowering hardy trees and shrubs. However, few of them tolerate dry climates or alkaline soil. The purple-leaf plum is an exception.

Discovered in the garden of the shah of Iran in the 1800s by a Frenchman, it flowers as spectacularly in Tucson as it does in Boston. The mass

of pale pink flowers, which appear in early spring on leafless branches, are succeeded by reddish purple, oval, pointed leaves.

Plants grow to 30 feet tall with a dense, billowing habit. They look particularly attractive beside trees with light green leaves, such as weeping willows. Transplant them from bare-root, containerized, or balled-and-burlapped stock into any fertile, well-drained soil in full sun. Use them as a lawn highlight, as a tall accent, or to create leafy avenues.

## Pyrus calleryana
### Bradford pear

The tree commonly sold as Bradford pear—actually a catchall term describing several varieties of *Pyrus calleryana*—began to gain prominence in recent years at the U.S. National Arboretum, where director John Creech had planted a number of specimens collected in northern China. He spotted one among them that outshone all the rest, with a billowing pointed shape, multitudes of extraearly spring blossoms, and dazzling fall color. He propa-

gated it by cuttings and introduced it to the public through the nursery industry.

Bradford pear is so earlyflowering that it often blooms ahead of forsythia, lighting up bleak landscapes with a beacon of glistening white, crabapplelike blossoms. These are followed by decorative, heartshaped green leaves and inconspicuous, inedible fruit. In autumn, leaves turn shades of yellow, orange, and red. Plants grow to 50 feet high.

Bradford pear thrives from northern Florida into Canada on the Atlantic Coast, and from San Diego to Vancouver, British Columbia, on the Pacific Coast. With the exception of desert-dry and boggy ground, Bradford pear tolerates a wide range of soils. It is highly pollution tolerant and is widely planted as a street tree. It is also useful as a lawn highlight.

Many mail-order nurseries offer bare-root cuttings, which are economical for mass plantings, but try local sources for more reliable container plants. The cultivar 'Aristocrat' has a more spreading habit and is more resistant to branch and trunk damage than other types.

*Raphiolepis umbellata*

*Rhododendron* hybrid 'Stewartstonian'

*Rosa rugosa*

### *Raphiolepis umbellata*
Indian-hawthorn

In the climates where Indian-hawthorn is hardy—the South, Southwest, and West Coast—no other shrub will provide such a dense flowering display under hardship. Provided that drainage is good, Indian-hawthorn will tolerate poor soil, air pollution, and drought. Its mass of white or pale pink clustered flowers appear at least twice a year, usually in early spring and autumn. These make a striking contrast with the lustrous, dark green, leathery leaves.

Plants grow 4 to 10 feet high depending on variety. 'Enchantress' is a lovely, compact plant with rose pink flowers, suitable for pruning with shears to create low mounds and hedges. 'Majestic Beauty' is free flowering and grows like a tree to 10 feet; it is useful as a lawn highlight or tall accent along a foundation.

Buy Indian-hawthorn in containers at the nursery. Do not confuse it with English hawthorn or Washington hawthorn, which are much hardier, belong to a different genus (*Crataegus*), and prefer cooler climates.

### *Rhododendron hybrid*
'Stewartstonian' azalea

Azaleas and rhododendrons are both members of a plant genus called *Rhododendron,* but they differ in significant ways. Generally, azaleas are bushier and more compact in habit, and they have smaller leaves and smaller flowers (but more of them) than the plants commonly called rhododendrons. More importantly, azaleas are much easier to grow, tolerating heat and drought more successfully.

One of the easiest-to-grow azaleas owes its existence to Joseph B. Gable (1886–1972), a Pennsylvania farmer who spent a quiet lifetime creating some of America's finest hybrid rhododendrons and azaleas. Azalea 'Stewartstonian', introduced in 1952 and named for Gable's home town of Stewartstown, Pennsylvania, is his most famous innovation. Exceptionally cold and heat tolerant, it flowers well where many other azaleas will perish. Each spring it is covered with bright red, 2-inch-wide, star-shaped flowers that almost smother the plant. In fall its narrow, oval, evergreen

leaves also turn brilliant red. Plants grow slowly, to a height of 15 feet and an unusually slender width of 6 feet.

Use it along a house foundation, in shrub borders, scattered through woodland, along stream banks, massed in shade, or in sun at the edge of a lawn. Plants look especially vivid in front of pines or other needle evergreens. They also make a fine flowering hedge.

Azaleas will not grow everywhere, of course, and this one is no exception. Like its more temperamental cousins, it prefers light shade and highly acid loam soil rich in humus, especially in the form of peat, compost, or leaf mold; a mulch of bark chips also helps conserve soil moisture. Its preference for highly acid soil and protection from wind makes it a poor candidate for hot, dry, blustery regions with alkaline soil, such as the Great Plains. However, it is well worth trying if you can give it what it needs.

### *Rosa rugosa*
Rugosa rose

Most roses seem to have only one purpose in life—to flower profusely no matter how much

neglect they receive. Indeed, about the only condition that will diminish their flowering is inadequate light. Provide rugosa rose with full sun and good drainage, and it will flower throughout the summer, climaxing in June and July. It ends the season with a crop of decorative orange-red rose hips.

Rugosa rose is one of many old-fashioned species roses that are hardier, more disease resistant, and more carefree than the modern hybrids that have succeeded them. A valuable shrub for difficult soils, including sandy soil and soil exposed to salt spray, rugosa rose makes a good windbreak in coastal gardens. Purchase plants in bare-root form by mail or in containers from local nurseries. Plants grow quickly to 6 feet in a dense mass of stout, thorny canes with lustrous, dark green serrated leaves. The 3- to 4-inch blooms are available in rose, purple, or white, and in single and double varieties.

For mild-winter climates, other good species roses include Lady Banks's rose (*R. banksiae*), which bears

*Syringa vulgaris*

*Vinca minor*

*Wisteria floribunda*

a great quantity of double yellow flower clusters on billowing, climbing plants; and cherokee rose (*R. laevigata*), which has 4-inch single white flowers on canes that can climb 20 feet high. A hardier beauty for cold winters is Father Hugo rose (*R. hugonis*), whose 3-inch single yellow flowers bloom from bushy plants with cascading canes.

### *Syringa vulgaris*
Common lilac

Although the common lilac does not do well in southern states, it is practically indestructible in the North. Each spring this dense, bushy shrub, which grows up to 10 feet high, produces an abundance of fragrant flower clusters in blue, purple, white, or red. When plants become tall and sparse flowering, cut the entire plant to within 12 inches of the soil immediately after flowering and it will sprout new growth. Fertilize with a balanced granular fertilizer or with well-decomposed animal manure to promote generous flowering.

Plants tolerate poor soil if drainage is good. They demand full sun. Although lilacs are

susceptible to powdery mildew (a gray mold that slightly discolors the leaves in summer), this occurs after flowering and rarely harms the plant.

Use lilacs as a lawn highlight, along a house foundation, as a screen or hedge, or as a source of cut flowers. Hybridizing has produced some spectacular large-flowered kinds, notably 'Marechal Lannes', with deep purple flower clusters up to 12 inches long. Several dwarf species make good highlights in perennial borders, particularly *S. microphylla* (little leaf lilac).

### *Vinca minor*
Periwinkle

If you have a patch of shade where grass will never grow, don't plant pachysandra or English ivy—the most common ground covers for shade—until you have tried *Vinca minor* or its more heat-tolerant (but cold-sensitive) southern relative, *Vinca major*. Besides providing a beautiful evergreen carpet of dark green, glossy oval leaves, both species flower continuously for several months beginning in early spring, displaying masses of

star-shaped blue flowers. *V. minor* flowers are ¾ inch wide; *V. major* flowers are slightly larger. A white-flowering variety is also available. Plants grow 4 to 6 inches high, hugging the ground in a mat.

Periwinkles will also thrive in sunny locations. They tolerate poor soil, but perform best in moist, humus-rich loam. In addition to providing a durable, attractive ground cover for shaded places, periwinkles can be used as an edging for paths, or to cascade over the edges of tubs and window boxes.

Mail-order sources supply bare-root cuttings; local nurseries offer transplants in packs of six or more. Established clumps can be divided to make new plants: In spring or fall, separate the roots and plant them in new locations, 12 inches apart.

### *Wisteria floribunda*
Japanese wisteria

You can travel the length and breadth of North America and find Japanese wisteria vines flowering as beautifully in Quebec as they do in California. There are years, admit-

tedly, when a late freeze in the north may nip the buds and prevent a good display, but wisteria is a hardy favorite almost everywhere. It scrambles up walls, fences, or arbors to dangle beautiful foot-long clusters of sweetly scented blue, white, or purple pea flowers.

Plants grow astonishingly fast—up to 10 feet a year—and reach heights of 100 feet or more. The narrow, bright green, pointed leaves usually appear after flowering ends.

Wisteria is easy to plant from containers. It prefers fertile, well-drained loam soil in full sun. In phosphorus-poor soil, adding superphosphate or bonemeal will promote extra-heavy flowering.

Use wisteria as a carefree vine to cover monotonous expanses of wall or fence, or to drape elegantly over arbors or pergolas. Keep plants bushy and compact with hard pruning of the branch tips after spring flowering. Tree-form wisterias, which have been pruned by the nursery to make a straight, self-supporting main trunk and a rounded foliage canopy, can be purchased in balled-and-burlapped form for use as a lawn highlight.

## GLOSSARY

*If you're new to gardening, you may be unfamiliar with some of the terminology you'll encounter when shopping for plants and supplies. Below are definitions of the gardening terms used in this book.*

**Acid, alkaline** Two extremes of chemical reactivity in soil, as measured by the concentration of hydrogen ions (pH). Acid soil (low pH) is found mostly in forested areas with high rainfall; alkaline soil (high pH) is found in desert areas. Some plants favor acid soil, some alkaline.

**Annual** A plant that blooms, sets seed, and dies within a single year.

**Back of the border** The rear of a flower border, where tall plants look best.

**Balled and burlapped** A form in which some trees and shrubs are sold. Balled-and-burlapped plants have been dug up with their roots encased in a sizable ball of soil, then wrapped with burlap to keep the rootball intact.

**Bare root** A form in which some trees and shrubs are sold. Soil has been washed from the roots and the roots kept alive in a wrapping of moist sawdust, peat, or shredded newspaper. Mail-order plants are often shipped in this form.

**Bed** An island of soil for planting, usually surrounded by lawn, brick, or flagstone.

**Bedding plants** Ready-grown annuals for transplanting, usually sold in six-packs.

**Bicolor** A flower whose petals have two colors, either together on a single petal or on adjoining petals.

**Biennial** A plant that completes its life cycle within two years, blooming and dying during the second year.

**Border** A narrow planting area along a wall, fence, path, or hedge.

**Bulb** A swollen underground section of stem from which some plants, such as tulips, regenerate.

**Compost** Decomposed plant or animal material used to fertilize and condition the soil.

**Corm** A scaly, bulblike underground stem from which some plants, such as crocuses, can reproduce.

**Cultivar** Short for "cultivated variety." A distinct plant, produced in cultivation, that differs from wild varieties; for example, the 'Queen Sophia' cultivar of *Tagetes patula* (French marigold).

**Cultivate** To turn over and break up the soil so that plants can grow freely in it.

**Deadhead** To remove faded blooms from a plant in order to stimulate flowering and prevent seed formation.

**Double flowered** Having more than one row of petals. A semidouble flower has two rows; a fully double flower has a mass of petals, producing a ball shape.

**Drift** An informal mass of flowers planted along a hillside or in a woodland or meadow.

**Eye** The center of a flower, where the petals meet. Some plants have eyes that contrast with the petal color; for example, black-eyed-susan.

**Flats** Shallow planting boxes of wood or plastic.

**Foliar feeding** A method of fertilizing plants by spraying the leaves with liquid fertilizer.

**Forcing** A method of inducing flowering plants to bloom early, usually by taking them indoors during winter and providing light and warm temperatures.

**Foundation plant** A plant that is especially suitable for planting around a house foundation, such as azalea.

**Free flowering** Producing an unusually large quantity of flowers over a long period. Triploid hybrid marigolds, for example, are free flowering.

**Genus** A group of related species.

**Habit** The characteristic growth pattern of a plant.

**Hardy** Able to withstand frost or survive freezing weather.

**Herbaceous** Having stems that are soft and green rather than woody.

**High-nitrogen fertilizer** Fertilizer that has a high nitrogen content, such as dehydrated steer manure.

**High-phosphorus fertilizer** Fertilizer with a high phosphorus content, such as bonemeal, used mostly to stimulate flowering.

**Humus** Beneficial organic material in the soil, normally produced from decaying animal and vegetable matter. Good sources of humus include compost, peat, leaf mold, and animal manure.

**Hybrid** Offspring of two plants of different species that do not usually mate in nature. Hybrids usually have desirable traits, such as disease resistance, that are superior to those of the parents.

**Lawn highlight** A plant that is sufficiently ornamental to be pleasing when set out alone in the middle of a lawn.

**Lead shoot** The stem that is the plant's primary point of growth. When this is cut or injured, the plant will produce side branches, creating a bushier growth habit.

**Mulch** A soil covering applied to retain moisture, deter weeds, and stabilize soil temperature; for example, straw, shredded leaves, wood chips, or black plastic.

**Offset** A young plant produced at the end of an underground stem. Yucca is an example of a plant that reproduces by offsets.

**Perennial** A plant that lives more than one year.

**Pinch** To cut off the growing tips of plants to make them produce side branches, usually by using the fingernails of thumb and forefinger.

**Prune** To cut off parts of a plant using a pruning tool, such as shears.

**Rhizome** A swollen underground stem capable of producing a new plant. Irises, for example, spread by rhizomes.

**Shear** To cut a mass of twigs or flowering stems with pruning shears so that the stems are all the same height.

**Single flowered** Having flowers with a solitary row of petals.

**Soil conditioner** A substance such as peat moss or compost that is added to the soil to improve its texture, to neutralize acidity or alkalinity, or to supply nutrients. Also called a soil amendment.

**Species** A unique plant occurring in the wild.

**Tender** Susceptible to damage by frost.

**Throat** The tubular wall of a trumpet-shaped flower such as foxglove.

**Tip-prune** To cut off the tips of branches, usually in order to make a tidy shape or to promote a second flush of flowers.

**Variety** A plant that is distinctly different from the species to which it belongs; for example, in color or height.

# MAIL-ORDER SOURCES

If you can't find the plants you want locally, try these mail-order sources for items described in the text.

## SEEDS

**W. Atlee Burpee Seed Co.**
300 Park Avenue
Warminster, PA 18974

**Comstock, Ferre, and Co.**
263 Main Street
Wethersfield, CT 06109

**Henry Field Seed and Nursery**
Shenandoah, IA 51602

**Guerney Seed and Nursery**
Yankton, SD 57079

**Hastings**
1036 White Street, S.W.
Box 115535
Atlanta, GA 30310-8535

**J. W. Jung Seed Co.**
335 South High Street
Randolph, WI 53957-0001

**Earl May Seed and Nursery Co.**
208 North Elm Street
Shenandoah, IA 51603

**Park Seed Co.**
Cokesbury Road
Greenwood, SC 29647-0001

**Stokes Seed Co.**
1278 Stokes Building
Box 548
Buffalo, NY 14240

**Thompson and Morgan, Inc.**
Box 1308
Jackson, NJ 08527

**Otis S. Twilley Seed Co.**
Box T65
Trevose, PA 19047

## BULBS

**Breck's Bulbs**
U.S. Reservation Center
6523 North Galena Road
Peoria, IL 61632

**Peter de Jager Bulb Co.**
188 Asbury Street
Box 2010
South Hamilton, MA 01982

**Michigan Bulb Co.**
1950 Waldorf
Grand Rapids, MI 49550

**Rex Bulb Farms**
Box 774
Port Townsend, WA 98368

**Swan Island Dahlias**
Box 700
Canby, OR 97013

**Van Bourgondien Bros.**
Box A
245 Farmingdale Road
Route 109
Babylon, NY 11702

## LIVE PERENNIALS, TREES AND SHRUBS

**Bluestone Perennials**
7211 Middle Ridge Road
Madison, OH 44057

**California Nursery Co.**
Box 2278
Fremont, CA 94536

**Emlong Nursery, Inc.**
Box 236
Stevensville, MI 49127-0236

**Girard Nursery**
Box 428
Geneva, OH 44041

**Kelly Nursery**
Box 800
Dansville, NY 14437

**Lakeland Nursery**
Sales Dept. LKL 2662
Hanover, PA 17333

**Mellinger's Nursery**
2310 West South Range Road
North Lima, OH 44452-9731

**J. E. Miller Nursery**
5060 West Lake Road
Canandaigua, NY 14424

**Savage Farm Nursery**
Box 125
McMinnville, TN 37110

**Springbrook Gardens**
6776 Heisley Road
Box 388
Mentor, OH 44061

**Spring Hill Nurseries**
6523 North Galena Road
Box 1758
Peoria, IL 61656

**Spruce Brook Landscaping**
Route 118, Box 925
Litchfield, CT 06759

**Stark Bros. Nursery**
Highway 54 West
Louisiana, MO 63353

**Andre Viette Farm and Nursery**
Route 1, Box 16
Fishersville, VA 22939

**Wayside Gardens**
1 Garden Lane
Hodges, SC 29695-0001

**White Flower Farm**
Route 63
Litchfield, CT 06759

## U.S. Measure and Metric Measure Conversion Chart

| | | Formulas for Exact Measures | | | Rounded Measures for Quick Reference | | |
|---|---|---|---|---|---|---|---|
| | Symbol | When you know: | Multiply by: | To find: | | | |
| Mass (Weight) | oz | ounces | 28.35 | grams | 1 oz | | = 30 g |
| | lb | pounds | 0.45 | kilograms | 4 oz | | = 115 g |
| | g | grams | 0.035 | ounces | 8 oz | | = 225 g |
| | kg | kilograms | 2.2 | pounds | 16 oz | = 1 lb | = 450 g |
| | | | | | 32 oz | = 2 lb | = 900 g |
| | | | | | 36 oz | = 2¼ lb | = 1000g (1 kg) |
| Volume | pt | pints | 0.47 | liters | 1 c | | = 8 oz | = 250 ml |
| | qt | quarts | 0.95 | liters | 2 c (1 pt) | = 16 oz | = 500 ml |
| | gal | gallons | 3.785 | liters | 4 c (1 qt) | = 32 oz | = 1 liter |
| | ml | milliliters | 0.034 | fluid ounces | 4 qt (1 gal) | = 128 oz | = 3¾ liter |
| Length | in. | inches | 2.54 | centimeters | ⅜ in. | | = 1 cm |
| | ft | feet | 30.48 | centimeters | 1 in. | | = 2.5 cm |
| | yd | yards | 0.9144 | meters | 2 in. | | = 5 cm |
| | mi | miles | 1.609 | kilometers | 2½ in. | | = 6.5 cm |
| | km | kilometers | 0.621 | miles | 12 in. (1 ft) | | = 30 cm |
| | m | meters | 1.094 | yards | 1 yd | | = 90 cm |
| | cm | centimeters | 0.39 | inches | 100 ft | | = 30 m |
| | | | | | 1 mi | | = 1.6 km |
| Temperature | °F | Fahrenheit | ⅝ (after subtracting 32) | Celsius | 32°F | | = 0°C |
| | °C | Celsius | ⅝ (then add 32) | Fahrenheit | 212°F | | = 100°C |
| Area | in.² | square inches | 6.452 | square centimeters | 1 in.² | | = 6.5 cm² |
| | ft² | square feet | 929.0 | square centimeters | 1 ft² | | = 930 cm² |
| | yd² | square yards | 8361.0 | square centimeters | 1 yd² | | = 8360 cm² |
| | a. | acres | 0.4047 | hectares | 1 a. | | = 4050 m² |

# INDEX

*Note: Page numbers in boldface type indicate principal references; page numbers in italic type indicate references to illustrations.*